The FTL Papers

Book Five of the Urbis Phobia Series by Tomas London

Published in 2015 by Amazon Inc.

Isbn: 978-197-355-8224

Table of Contents:

1. The FTL Papers as created. November 1983. Page 3
2. Our Galaxy As It Really Is, 2015. Essay. Page 25
3. The Year 2290, essay about climate change from 2015. Page 31
4. The Flavian Philosophy, data from fiction. Page 38
5. Atlantis Essay, data from 2005. Page 40
6. Fuck Atlantis Redux. Shocker from 2015. Page 86
7. Cydonia My Love, short fiction. Page 95
8. Rosetta Stone, fiction from 2002. Page 108
9. Rain, very short fiction from 2012. Page 118
10. Fear City, short fiction from Book One. Page 130

The FTL Report of 1983

Introduction:

Hypothesis on a plausible method of "Faster Than Light" travel by Tomas London. Conceived in November 1983 and rewritten in 2015. Please keep in mind that this is only collection of many points of view from the minds of experts in the fields of Physics, Math and Engineering and thus synthetic. In other words: The author does not present himself as an expert in those fields.

The basic premise of the Theory of Relativity is that the speed of light, "c" is an absolute barrier which neither matter nor energy can break through. "C" seems to be the highest possible velocity but that may be refutable, as the Addendum shows. Another vital premise is that Time will decrease to absolute stasis as an object of matter accelerates to the speed of light.

The above premises are the argument against FTL, at least as far as manned flight is concerned. It goes without saying that the Math behind a certain Theory Of Relativity is beyond reproach and therefore provides a solid foundation for further Physics research. In terms of explaining motion within the Universe; it must seem to casual observers that the thinking of Einstein has replaced earlier laws of relative motion (ie. Newton). This has been a trend apparent over the many centuries as velocities of such moving objects have increased. Yet, this trend is only apparent... for there may be no disparity between the Laws of Newton and those of Einstein. Our reasons for this will be explained in detail later.

For now, any major deductions which others tend to make from Einstein theories, whether General or Specific, are held to be against any notion of FTL. As stated above, the literature we have so far usually claims that some sort of "barrier" exists at "C" which prevents any spacecraft from going over it. This barrier of 186,200 miles per second must be a semantic one, like the sound barrier before the time of Yeager. Then, we must deal with relativistic velocities per se. The problem of **momentum,** or the end result of increasing velocity (plus time dilation and increased mass, density plus length of ship) is what we will discuss later on in this essay. For now we will discuss simpler notions of relative motion.

The basics of motion are that an object (such as a photon of very ordinary light) will obey the Newtonian laws (such as those of action reaction or the Law of Conservation of Energy) and thus be fairly easy to understand as universal laws. Our newer theories of the 20th Century add to them, not refute them. That is my contention. To use the photon as example of this rather complex train of thought, just consider some facts as far as we know them: a photon leaves a star and then travels through Space forever. According to both

Newton and Einstein, this photon will continue on its path without any change in velocity unless there are changes in gravity and medium density along its path (in other words in perfect vacuum with no interfering gravity wells). This goes only for one particle, regardless of wavelength.

The energy needed for the photon to leave the star was high enough to boost it out of a very strong gravity well, and can be expressed as the famous equation "E=mc2" resulting in a standard velocity. As long as a perfect vacuum and steady gravity are kept up, this velocity (c) will remain the same as it was at the very point in time this photon first left the surface of its star. This photon will thus travel at a steady 186,200 MPS forever. Yet since it is only a particle it has no energy other than the amount it had as it first left the star's surface or, in other words a certain amount coming from the reaction within the star itself.

So since Space contains obstacles such as gas, dust and solid objects with their own gravity, the direction and velocity of our sample photon will soon change, which means that "C" can indeed be variable. The point is that all of this had been taken into consideration by Einstein in his books up to 1956, thus his theories are not as simple as they seem. To summarize: Under normal conditions the speed of light is not constant at all.

In the early years of this century, speculations on reactions between particles and various fields have led to such theories as the following: Einstein's Four Fields plus his UFT/Unified Field Theory) or Lorentz with his "wave mechanics". However, such things belong in another essay. The point of presenting a sample particle is this: my photon has no way of altering its velocity on its own. It is like a bullet shot out of a rifle. But a spaceship has its own fuel and reaction drive, which enables it to gain momentum. We can assume that its engine must be some sort of rapid Plasma Drive and obviously powered by nukes. Our real problem, as we encounter it in FTL is: How to deal with relativistic effects as they arise in flight.

So recall that a travelling particle such as a photon can add nothing nor subtract from itself. Its momentum comes from the source itself. Yet, a ship has its fuel supply which is matter (mass) lost and therefore energy gained. Velocity can be added to. We must assume that Newton's laws will apply to our ship until "C" is attained, so that by then Relativity takes over totally. Finally, the laws of Newton plus our known Laws of Physics are suspended. That will come into play due to the slowing down of Time itself.

We must recall at this point, that every moving object in the universe is within its own "frame of reference". Any steady particle moving along at exactly "C" is not at all connected to its original source of energy (such as a distant star or lamp or even a campfire). This may explain why emissions of waves from other stars (such as TV or radio signals) have not yet been detected by Project Ozma / SETI. In fact, we have mostly given up on them. Ordinary light spreads with distance and there is a lot of dark matter out there to absorb it.

A ship travelling from our star system (Sol) to another will fly relative to where it came from. It has its own frame of reference in time and space.

Relative to Sol it then must speed up until it gains the velocity "C". At that point it must reach infinite velocity, mass and size. Time will slow down to zero compared to what it was at start of our trip. Obviously, "time" itself is an abstraction. So the only things that can "slow down" on board any relativistic vessel are the physical processes in and of themselves: subatomic motions of particles, molecular motion, chemical processes and then even more complex biological and mental ones. For example, a crew member floating across a room on it will move more slowly than he would on a parked or motionless ship. But of course, he will not perceive this fact. To him, time will flow along; by any standard; at the same apparent rate as it does on Earth or on any planet which is motionless when compared to the ship. To be concise: the passage of time as it is felt by any person on a ship travelling at relativistic speeds will be the same to him as it always was and is, period.

Why did I say that? Well there is an idea in Physics which says that the ship will eventually reach "Stasis" or a state of complete stop of "Time". That alone should stop our ship from accelerating any further. It is as if the pilot cannot physically press the button to make his drive increase thrust, or as if the fuel could not physically eject itself out of the rear nozzles (remember that we are talking plasma rives here, not Dean Drives...). Clearly that idea was not yet proven since we have not yet gone more than 100,000 miles per hour in Space. It is also said that Time Dilation is immense as the ship approaches "C" which leads to many stories about this very topic in fiction. The above process may exist but it will only apply as long as our ship in question travels at any velocity (relative to its point of origin, in this case Earth) from circa 10% of C to 99.999%; ad infinitum. But what will happen to our ship in a practical sense if it exceeds "c" which is actually possible?

To answer that question we need to go to the findings of H.A. Lorentz and Gerald Feinberg. Lorentz dealt with wave mechanics and then developed his rather obscure Lorentz Curve as result while Feinberg more recently came up with his famous "tachyon" which is still a very hypothetical particle. Unlike antimatter and neutrinos, tachyons have not been proven. We may not even need to have proof of their existence for our FTL hypothesis, so tachyons will not be our topic. Nevermore.

However, we shall soon see that the Lorentz Curve can be more relevant. The author has by now extrapolated new and never before seen FTL Curves from it as we will see on our diagram. Note: In 1983, as this essay was written in its original form and without knowing anything of Lorentz, such a curve was original as a new concept. But its source was unknown. According to conventional thinking, it must then be physically impossible to accelerate a ship beyond "c". Our ship is near stasis, so that the entire mass of the ship is losing energy. As "c" approaches, subatomic activity slows down ever more until all matter in the ship attains a total lack of energy, which is Stasis. None of this will be noticeable to our crew because they are still in the same frame of reference.

We must ask ourselves what this state of non - motion, or very specifically put; Stasis, implies. This may lead to pure speculation but let us pause to deduce the nature of matter and/or energy as it exists in "stasis". By the way, this state was induced only by such relativistic effects themselves, as they already exist on any ship as soon as its up there in that condition. It is not caused by pressing some strange button or other science fiction device. This is where some glib, flippant and halfbaked concepts in fiction, such as say hyperspace or wormholes come from. They are okay as labels for my findings in FTL, but were never backed up by any empirical proof. Bogus labels are not needed in this age. That point was made by John W. Macvey.

[Pause: What happens after Stasis?]

Nothing has actually happened to the crew members after that soi disant "light barrier" was crossed as long as we consider only their own frame of reference. After all, time goes on for them as before. Fuel continues to be consumed energy lost and their velocity increased. Only a tiny thrust of reaction drive was ever made to allow acceleration itself to continue past "c" so the extra momentum or any change, was not perceived by them.

This brings us back to the old sound barrier, which was once considered to be substantial. Yet it was and is only an event horizon, and chances are good that the light barrier may prove to be the same. One simply cannot hear the sound of jet engines as they pass a certain velocity (circa 600 mph) because now the plane outruns the observer's sense of hearing. Likewise, we cannot see any FTL ship as it approaches or recedes from us because it was always totally out of range of vision. Yet none of this implies any physical barrier.

Our key point here is that matter in stasis due to relativistic effects may have properties that differ from those of matter as it exists under normal circumstances. After all, where has the energy gone which used to be in the matter? The Law of Conservation of Energy tells us that energy is never destroyed but converted into something else. Any ship in stasis is simply Mass totally devoid of Energy. So where has the energy gone? Not into our reaction drive, as that loss has been accounted for by fuel consumption. Here is my speculation: the ship becomes a form of energy not matter. It follows that as soon as the ship attains "c" its entire mass becomes a form of energy not yet perceived... or better yet... plasma. It is interesting to speculate whether this plasma expands or contracts as it crosses Space. This is what we expect in FTL. As we see it, any contraction with its molecular configuration maintained is the most likely alternative to consider. Will the ship collapse into a sort of miniature black hole? (Again, that will not be achieved by some bogus device of a mechanical nature but must be the natural end result of relativistic effects themselves.) This sort of endless compression of a ship's volume is another speculation which we will not get into here for the sake of brevity. However, it will allow the ship to build up great momentum. A ship could also expand out into the volume of Space after it passes "c" so that it will then maintain its mass but not its density. That will allow it to cross a vast distance of many light years. This sounds like a sort of "ghost ship" which is invisible and can pass through any "dark matter" found in Space.

But as we have said, a contraction of volume is more likely. It could also be called an implosion. Since the total absence of energy must make the particles of a ship implode (attract each other) then it seems likely that the ship will contract until it reaches "c" and then implode even further as a singularity. We assume that the configuration is maintained inside, so that a pilot or timer device can halt the process which begins a sort of deceleration. As soon as that happens, it will expand outward. By the time a ship slows down to "c" at the other end of its trip, it will be its normal size again. Mass should then be the same except for a minor loss in fuel. The reason for the above three paragraphs is this: there is an old and popular assumption that a ship must explode, implode or otherwise get destroyed as it reaches "c". However, that may apply only to ships in a normal frame of reference (ie. relativistic). Anything in Stasis must be very stable, and thus be able to maintain its atomic configuration in spite of any given changes in volume and/or density. At this point our theories become spooky.

Total stasis poses an obvious problem in that if everything on ship is in stasis, how can the drive be reversed, whether by a person at its controls or a timer device? Would not everything on board be frozen in time? Motionless forever? The reader may have thought of that by now. Actually, that infamous moment in time when "c" is attained (along with total stasis) is so tiny and over so fast that it is not measureable. Being a mere instant in time, it surely does not count as much as we once thought.

Our ship will have just enough energy left to push its velocity past "c" which will then start the process of time reversal at once and automatically. Now we have finally let the cat out of the bag (as the old saying goes) by mentioning the newer concept of Time Reversal. This is the one that has led some of the newest innovators (such as Feinberg) to invent concepts like "tachyons". (They may exist but are too complex as topics to get into.) As soon as that occurs, any sort of Stasis is no longer operating, which means all bets are off. But why think that time should reverse at all after "c" has been passed? For that, Einstein has an answer. We know that time slows down within the frame of reference of our ship. If we follow that logic, then we should soon conclude that time will start to flow backwards at once... but since we also know that Time is an abstraction (according to Einstein) then we must conclude that something else will occur. So instead of Time on the ship "reversing" which is logically not possible, we can more easily have Space being somehow affected in a negative way or "annihilated". This is where time dilation comes in. Stasis has been saving our passengers from the damage of gee forces. Then at "c" all existing stasis effects drop off, allowing our pilot to alter thrust at some point in FTL deemed to be midpoint for his trip. **OK? So, to wit:** To be successful in astrogation, the exact point of center of distance between two stars must be found and calibrated.

[Finally, a refreshing pause for reflection. No more juvenile BS, as you readers may find in our sensation oriented fiction.]

Of course, we have no such instruments today. The two frames of reference (the outside universe as compared to the small one inside our ship) never meet in any way. It has been calculated often that a ship might be accelerated up to "c" within one month. We must use fission or fusion drives for this, certainly not chemical rockets nor some fictional hyper drive. Thus our interstellar voyage will take at least two months for its engines to speed up to "c" and then slow down to "zero" fully. Any time travel due to relativistic effects will be minimal because very little time will be spent in that zone either way. **Okay:** A recent formula comparing ship's time spent within very high sublight velocity to actual elapsed time outside has shown that it is a ratio of 1/70 at 99.999%. Therefore conditions would have to be extremely relativistic in order to get significant results. In other words, the idea in FTL is to boost speed as quickly as possible to avoid being trapped in Time and so, land on a star years too late. Time travel must be avoided; yet so must the dangers to human life due to high gee forces.

As we have said, time travel due to relativistic effects will be minimal, especially since star systems separated by many light years are wildly different frames of reference. The only part that may, in any academic way be debated, is the FTL part, so that it must be only Instantaneous. In other words, our ship spends one month powering up to "c" then traverses a distance of many light years in an instant, and then spends one month in its braking maneuver to come down from "c". We can just Go With The Flow. Like in 1968. This can be done by the use of simple retro rockets. However, the FTL section of our trip will be the longest in terms of Anything and also the shortest in terms of Real Time, which is also computer industry lingo. It will be the actual manifestation of the "process of time reversal" as "annihilation of space" or better yet, why not just say "teleportation"?

[**Note:** My one month estimate for up/down boosting is based on the fact that the human body can survive (at least with cushions and lying in a horizontal position) up to 30 gees of acceleration. That was a ball park estimate, but teleportation takes microseconds.]

Conclusion:

This essay may have been profound when it first came out in November of 1983, but it had its faults. We have since altered it to improve its logic. So it is now easier to read and has many new discoveries to back it up. Such as real antimatter, better fusion engines & even new planets around other stars. My own term, Annihilation Of Space as a complex way of saying "To teleport a ship across space" has been preserved. This serves to overcome a paradox (the one facing us after we pass C) and has been mentioned in a certain obscure film about the life of Steve Hawkins. This topic is usually referred to as Time and Space Paradoxes. To be frank, our common phrase, Annihilation of Space comes from an old novel by Ursula K. Leguin: The Dispossessed. Of course, as an author of only fiction, Tomas Londan, we feel that the best part of this essay was that **"S" Curve** of our original **1983** diagram. It has remained unaltered, and expresses any complex theories on teleporting which were borrowed from the experts. Such as aliens who did accost me personally on this Earth to communicate to me as a person, bizarre concepts of Space & Time. Nor do Sumerian ghosts and other ancient entities impress me. OK? So what gives?

The only reason I can offer this report is the history of our human sciences over the past century. Most of the major theories in physics were known by 1930. Then came a gap of 60 years until circa the Gulf War Era, by which time some great engineering feats had been accomplished: those second generation fusion engines that were being developed at the famous Lawrence Livermore Labs for UCB as well as antimatter found in major particle accelerators. Also we have the new Nonlinear Science coming out in the Nineties and promoted in its purest form by G. Nicolis of Belgium.

There is mystical thinking involved. Which was an issue of the 1980s Cosmology Fashion Set... as if they were selling glitzy dresses for Vogue Magazine... it crept into that first edition of my essay, but has been removed. We are now more pointed, detailed and coherent. There are quotes in Addendum below. Also we have now have a Bibliography. We found that words were not as useful in stating FTL reality as our diagram. It will be far better for reaching some understanding of this strange subject than mere semantics.

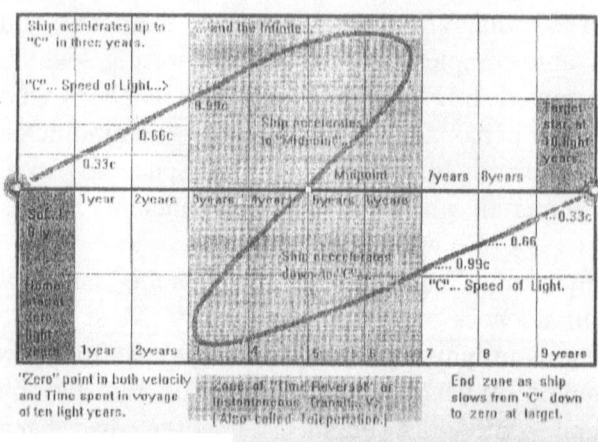

Addendum: Quotes

First: "In order to give physical significance to the concept of time, processes of some kind are required which enable relations to be established between different places." Einstein, page 28 of his book.

UFTL/ Main section ends/ Addendum starts here:

Short Guide To Early Physics

The essay, "The FTL Paper" was written in Nov. 1983 by Tomas Londan to point out possibilities in space travel. In this guide we quote a book by Henri Arzelies to further illustrate our theories.

We start with the Fitzgerald/Lorentz Contraction (p. 126) as in:

"Lorentz studies the basic problem much more intimately. The dimensions of a body are obviously related to the mean equilibrium conditions between the molecules of which the body is composed. Setting out from this idea, Lorentz demonstrates that the contraction can be deduced from hypotheses about the structure of the body and the intermolecular forces. He thus relates the phenomenon to a more general theory."

On the principles upon which Special Relativity is founded:

"... tries to deal with the idea generally, that they [meaning certain scientists] then obtain the Lorentz formula with a universal constant, independent of the properties of light, which replaces "C". In particular, they show that the principle of relativity would not be contradicted if the velocity of light did depend upon the velocity of the source."

And now the book (on page 123) as it relates to "Causal Links and the Temporal Order of Succession of Two Events; Time Reversal." So here goes:

"By reversing the velocity "V" we can reverse the order in which two events occur, if they have both occurred simultaneously but at different places in some Galilean system."

Then goes on with math that tries to prove out "time reversal" as a concept. Then goes on to wit:

"This is not possible. Whatever system is chosen, it is impossible to change the order of succession of two events which are (or could be) causally linked in some other system. It is sometimes said loosely that the Lorentz transformation conserves the future and the past."

That was on page 124. Another goodie:

"The Temporalization of Space or the Spatialization of Time...from the point of view of the relativistic formalism, the standard of length could equally well have been selected as the fundamental standard; relativistic thinking states only that the two standards are interrelated, and that there is but one fundamental standard. If however, the operational requirements are added, the standards of time seems to be the necessary choice. I know of no axiomatic construction which begins with the concept of length... if we allow the standard of time this primacy, we are led to speak of the temporalization of space."

That was pages 55-56. Also another source:

Quote: "Dialectical materialism recognizes practicality as the sole criterion of truth and regards a theory as economical when its accuracy has been confirmed in practice." (page 55) Karpov, 1951.

So what does all of this really mean? Our main theme was the "Lorentzian transformation " which to us is clearly a confirmation of our FTL Hypothesis. This came along with transformations, curves, contractions, waves and formulas in his books, which ran from 1892 to 1925. And those formulas become vital when they actually state such things as "v>c" openly. It was fascinating and is on page 277.

The following comes from page 277 of "Relativistic Kinematics" by Henri Arzelies:

APPENDIX ONE:

The Lorentz Transformation, which is employed when the relative velocities between reference systems are greater than "c".

A. General Remarks

On several occasions, we have envisaged the possibility of velocities greater than "c", measured with respect to a certain system of reference. Nevertheless, the velocity "v" which here appears in the Lorentz transformation has always been assumed to be smaller than "c"; and indeed up to the present time, experiment has never revealed the existence of higher velocities for the transport of matter or energy with respect to a Galilean system. In practice, the Lorentz formulae are of interest only for such Galilean systems which can be attained physically. There is of course the motion of galaxies to consider, but we have seen that this phenomenon is manifestly outside the domain of special relativity. In my present opinion, the restriction **v<c** is therefore of a purely empirical nature. It is, on the contrary, generally stated that the original **Lorentz Transformation** has no meaning for **v>c;** we shall show that this is too extreme an opinion. From the formal point of view, the constant **"c"** appears rather as a privileged velocity, a universal constant, than as limiting velocity.

[Except when Gee forces destroy our frail organic bodies. Better to send robot probes out beyond that nasty Kuiper Belt. So how about nuke thrusters at 50 gees? NASA and other agencies have talked about this before. Say something once, why say again?]

B. The Lorentz Transformation; use of imaginary x and y variables: Consider a system K1 with real variables x1 and t1. If we pass to a system K2, such that v>c, the variables x2 and t2 are imaginary. This is of no great importance if the systems K1 and K2 are given, and hence if we do not consider the transitional behavior between v=0 and v>c; x2 and t2, be they real or imaginary (with respect to x1 and t1 which are conventionally chosen to be real), provide a coherent reference system which we can apply to the study of such phenomena. The invariance of the propagation law is preserved. The transformation formulae for times (Section 63) and lengths (Section 65) become:

[Here comes a complex formula. We wanted to quote this verbatim, but lacked the Math symbols to do this on our PC. Can you do this on your scientific calculator, which you must have? We lacked modern symbols, which means that the average Terran has not gone beyond the Year 1905 in their thinking. OK? That was not our fault.]

Results are shown in Figs. 59 and 60. Since **v=c** corresponds to The Infinite or **s=0,** there is nothing to prevent us from representing the two types of behavior, **v<c** and **v>c,** on the same graph. Here comes the dramatic part. The climax, say. The following simple formula actually exists in the original book by Arzelies And must have been around even before 1966:

Verbatim: **"If v<c, then v>c."**

[End of quote. Are we happy?]

Addon: August 2016:

We had to dig for this file from old discs. It was lost in 2016 for some reason. Ghosts in our machine? So we have some new idea here. What inspired us? Back in '83, one winter day we had this idea: What if Starship shrank, and also got ever denser as it approached stasis? That will happen at "C". At that point we assume that we simply maintain configuration. (Atomic.) So then our pilot tries to make Ship accelerate to speeds greater than "C" which will create an effect that opposes our previous "shrink" process. As we get closer to target we have to decelerate just so we can land. Like, use retro rockets. This slowdown will cause Ship to expand. That will make it reach target faster. Density decreases. This will reverse time, which then, at that point, cause us to teleport. As quantum leap and/or paradox effect. We get a sort of weird tradeoff between Time & Space. Consult Zeno in fiction. The above was Earthian.

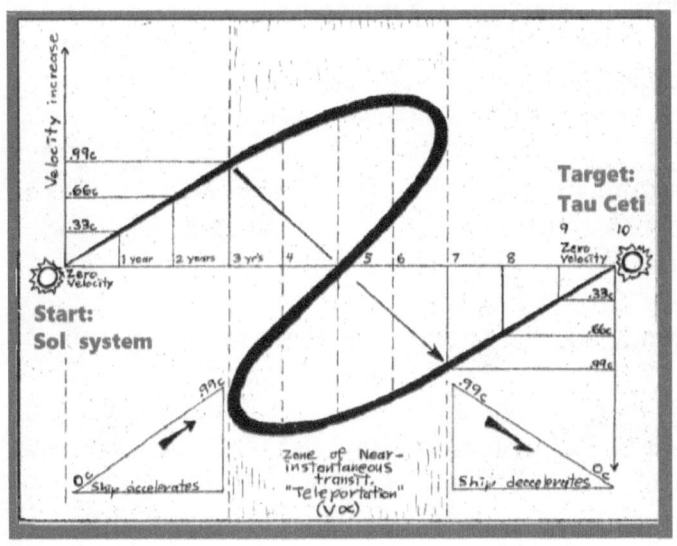

Bibliography

Arzelies, "Relativistic Kinematics" 1966 Oxford U. England. By Henri Arzelies, Pergamom Press. Call no. 531.1A698. Aleph System: 1175990

Bohm, "Special Theory of Relativity" David Bohm 1965 W.A. Benjamin, NYC. See quote on p.124

Einstein, The Meaning of Relativity. Fifth Edition. 1955 Princeton University Press. Princeton, New Jersey USA. Quote on p. 28 vital to essay.

Filmenovich, The Greatest Speed. 1983 Mir Publishers, Moscow, USSR. Quote p. 278-279

Johnston, "Biomed Results of Apollo" US Fed. Gov't 1975. By Richard Johnston with all relevant facts.

Leguin, Ursula K. The Disposessed This one is vital to my central idea, see page 275. Isbn: 0380003821 - Avon Hearst 1974 US

Lorentz, Henrik Antoon, see on Internet for all data.

Macvey, "Interstellar Travel, Past, Present and Future". 1977 Stein & Day, NYC, USA. John W. Macvey. Isbn: 038041368X

Mallove, Eugene, with Matlove, Greg, The Starflight Handbook. Isbn: 0471619124 - USA 1988 Wiley Co.

Nicogossian, Arnauld E. Space Physiology - 1982 US. Isbn: 0812111621

Nicolis G. U of Brussels, Intro to Nonlinear Science Cambridge U. Press 1995, Isbn: 0521462282

Walker, E.H. The Nature of Consciousness.
1970, Math Biosciences, p. 131-178.

Wolf, Fred Alan, "Star Wave: On Mind and Quantum Physics" by Fred Alan Wolf 1984 USA, Isbn: 0026308606

The FTL Papers ends here

November of 1983 by Tomas London.

Our Galaxy As It Really Is

Essay on Greenbank Formula in the Milky Way

Rules:

1. We assume there is no travel from one galaxy to another. Too distant for anything but observation, as we can tell from photos and other deep sky astronomy. Just consult your friends in STEMA.

2. No such thing as one big empire, of council ruling the entire Milky Way. Our galaxy is too large for that. Take the original Green Bank or Drake Formulas and then run them through your computer again. Add politics and ignore any homage to Hari Seldon, even what Cindi does.

3. We can call each star system an Empire with an actual Ruler and some organized culture. But how do they relate to each other? What do you think?

4. If any Alien claims to speak for some Emperor who "rules" this entire galaxy, then It is lying. That is a major point in the fiction of Tomas Londan. My

capitals.

5. IDIC as Vulcan saying is a simple way of assuming that the Green bank formula was one big step in the right direction. As in, there were many Races competing for this planet. In terms of exploring it. Some of them in the past may have tried to influence us. That is where media about **invasion** and **conquest** by aliens come from. They tend to be junk, but that is style. Hollywood. Some of them have sound basic ideas. We can call them Paradigms to be fancy. That means, be forgiving.

6. This activity did not work. Other races prevented any Invader type of Race from doing so. They may even be busy fighting each other Out There. Our wee planet is not important to them. They are more concerned with each other and we should think so. Celtic, eh? Day is long?

7. They have contacted us since 1947 in ships. Since 1977, most contact has been of the CE3 kind. In person. Without ships. Even here on busy streets in large urban areas. Teleporting is their method. That can be done without ships. So by all means read Passport to Magonia which has 923 cases. Dr. Vallee does not mention that famous and historical Heflin case of 1965. Why not? Because it was only a mere flyby. No aliens contacted. Yet it counts as a classic case. We expect some of you to compare the top scholars to match variables.

8. The Bible says that in times of war and rumors of war, some kind of Messengers From God shall descend from Heaven to help Man kind to survive. We assume. Often aliens have been taken to be such Saviors. As of 1945 that is so obvious. Ho hum.

9. Tat twam asi.

In the SF genre, there are indeed many examples of some Monster Empire ruling an entire galaxy. Well, not this one. Nor Messier 31. Too big. T.F.B. Was that a good theory? Asimov's Foundation series was a serious study. Next came Valerian with his Central Point. Democracy versus his Admiral. Next, we have Star Wars by George Lucas. At least that had an Emperor. Well, we now have a nice tradition in fiction. It is valid, but hard to create in fact. One more factoid: Laureline hates her boss. He is a dignified elderly man in a red tunic. Red. With lots of décor. We love it. She hates it. She hates him. Later on, in the Tran series, we see the same dude in his red tunic. Many medals. One problem: He is Evil. And a Man. Neat, eh? Now we know where Adz' degaraal comes from. At least fiction is consistent. We like tradition.

Here is what probably happened after the Time & Space Couple finally make an informed decision, then head back to Earth along with their old Admiral's fleet. (It is impressive.) Peace has been maintained at least for now. That was in 1976. Well, sooner or later, some other jerk alien might hatch notions of... galactic imperialism? Hmm? Maybe it will just be Zools. Or Ewoks? Or 61 Cygni? Somewhere out there in the Milky Way lie other freaks with their own fleets of warships. IDIC? That acronym goes Infinite Diversity In the Cosmos. We think. Eventually, as we have pointed out above, even in the Valerian Universe, which is fantasy and is okay for comic books, some logic must rule. Each System has its own Empire of planets and their moons, many of which you can live on. Plus vast clouds of gas which may contain resources. Billions of beings for citizens. If they can get out to some other system, conflict may start. That leads to the summer of 1977, it being the debut of Star Wars. By George Lucas. Which grew into a monster. (It made money, right?) So did Valerian, but only in Europe. War in space, as concept, for the first time revealed in modern style. It was not as corny as before. This whole Media storm of Space War helped to make the US space shuttle seem like a good idea. A few obscure writers in 1980 pointed out that the first basic models of this craft were only prototypes. Why not make them much bigger and then install nuclear engines? Chemical engines are just as dangerous and outdated. How can we explore the Solar System unless we take that next step? NASA could easily go nuclear.

They have money. Their nation, as in Congress & some voters, seem to be much too P.C. aka Politically Correct. Such sentiments are mostly partisan, folksy and not motivated by survival. Why bother? Let us have a Tea Party instead. The above is shameless propaganda. It was written in the Year 2018. Other writers of Amazon Inc. are into their own thing. Yet, what of my aliens? Are They up there laughing at us? Comments? Amen.

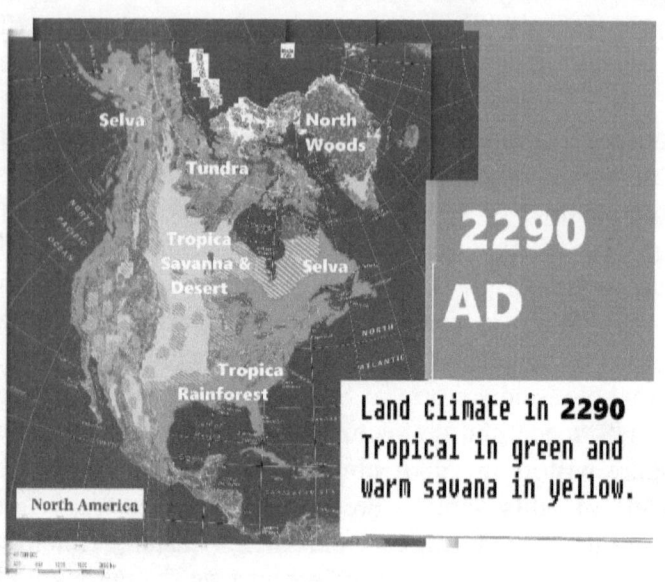

Land climate in **2290** Tropical in green and warm savana in yellow.

The Year 2290
And what it really is about.

It is now 2017. We used to hear a lot about the Apocalypse. It was usually taken to "be" or happen in or around the Year 2000. Or, according to a Mayan calendar, in 2012. That one began in 3114 BC, which makes "our" mystic date 5404 by their calendar. (That would be 6050 by the Judean one.) But what might be a real point in time for some End of the World event? We use astronomy and place it exactly in 2290 A.D. which is when the Age of Aquarius actually starts. Not "astrology" but modern astronomy. This fact must not be known to most of us. What can we deduce from this simple fact? A lot. Our first concept concerns Star Trek. That show was supposed to take place about that time, or just before it. So, in other words, the 23rd Century. By that time the following events seem likely to happen:

We may have faster than light travel. (FTL.) Plus quantum computers to go along with this, plus fusion engines for drives. We may also harness antimatter to create energy but only after we have run out of oil, gas and coal. To make it short, all of those resources will have run out on Earth by 2290. We will probably still have uranium, so we can use fission reactors down here, on other planets and on ships. We may find all of the above resources on Mars and some on various moons of our Solar System. We may have many large colonies, with families, on Mars and our own Moon, along with some smaller ones on the outer moons. Very likely the Jovian system will have permanent bases, and we will have landed on Pluto and Charon long before 2290, and by that, we mean manned missions. We will likely not encounter aliens until we leave our own star system. (There are some reasons for this, but they are hard to explain.)

By 2290, we should have starships similar to the NSS Enterprise, aka the old Federation Starship 1701. Now about the Greenhouse Effect. After all petro fuels have been used up, all ice covering Greenland, Antarctica and places like Baffin Island will be gone. We may then have forest, farms and cities there. We can certainly have mining there, and very likely find lots of oil, gas and coal. That will extend our life style and economy. However, by 2290 this planet may be much more crowded. Fuel has almost run out. At that point some kind of major war may finally break out, destroying most of our global culture. That will leave out colonies up there to house survivors. That will be the "real" Apocalypse. The difference is, our prediction is based on science. This may be the best incentive to explore other stars. As soon as we enter those systems, we must find that they are run by some kind of alien beings. That will be our greatest culture shock. We may even call our expanded human society a Federation of United Planets, but it would be arrogant to do so outside of that. Why? We suspect that very advanced aliens live out there, so they will not let us conquer or do anything beyond our own Kuiper Belt. In fact, that may start our first interstellar war, which we will lose. Right, James? As in Tiberius Kirk.

They may even force us into fighting each other, which may have been the main activity for most aliens who have been visiting our system, long before Mankind existed. We may be lucky to live on this planet, and be able to explore our own system. Probably, we will find eventually that some of those very aliens have been defending us, while others have been busy trying to enslave or destroy us. Thus, there was never any con census on our world, nor our kind. This theory gives rise to concepts of Good and Evil, which may have created many of our religions. The real source of Mono theism & also Dualism (ie. Cosmic) may have been this very simple concept which also inspired our first science fiction books. [Even better, see the last disk of Season Seven of the Ancient Aliens TV show, where in a V.I.P. meet of experts in Congress, in the Dee Cee, this very same point is made. They are cogent, and MUFON had a role to play. Date: 2013.]

We called ourselves the New Age as a joke. Predictions mentioned above were based on logical ideas of how and why society may or may eventually fall apart. That is what the Apocalypse really means. There was similar idea in our Holy Bible, but that was a narrow chain of events which had no relevance for our modern society. (That was about Rome being sacked in the Year 410 AD which has already happened, so is now ancient history.) Now we know why Christianity is almost dead, and why many of us have looked into some other religions (such as Mayan) for a new Judgement

Day, or Year of Global Retribution, or extreme climate change, or what not. This ancient, common search for collapse is an excuse for anarchy, war and revolution. We are just a bunch of hip people who grew up during the 1970s, who heard about some kind of "conspiracy" which sounded like bullshit at first, to be honest. That generation was too cynical to care about any revolution which happened 200 years ago in America and France. (The basic idea of Conspiracy was started only in the early 1800s by a few European radicals, and grew into a fad by 1968. By 1988 it was a joke.) Our group has been removing romanticism based on what media have been saying for 200 years. **Yet, by 2290, what you may call the Real Apocalypse shall likely happen!** These may be some astounding, yet coincidental events. Now, just for fun, you can still look up your Horoscope and compare two Signs of the Zodiac. See what makes Pisces so "lousy" and Aquarius so "nice" if anything. Each sign lasts for 2,160 years. Well you can assume that anything can happen within those long ages of our history. That is the hard part.

Note on new data from Internet:

This oft mentioned cycle of 25,920 years consisting of Ages of Zodiac of 2,160 years each. Well, it is usually called one Platonic Year. Did Plato invent or discover this? At least he was involved in our great Atlantis Mystery in many ways. Again, here is more proof. They also say that these swings, which start Ice Ages, first began in 2,700,000 B.C. So that we may assume that some large meteor hit Earth 2.7 million years ago. It may have crashed into Yucatan or Canada, forming the Gulf of Mexico or Hudson's Bay. That was long a major theory of climate change. Anyway, for that time, an Ice Age has come and gone every 30,000 and also 100,000 years. So these climate change estimates must be very rough. They do not match my precise astronomy cycles - my Platonic Years – with any reliability. But we can say, as a joke, that if we go by Platonic ways, the Year 2017 A.D. is now 21,327. The Year 2290 will be 21,600. And in my Book Two, 2084 will be 21,394. What is my point? Aliens may use this Platonic calendar. It should tie in with any Mayan or other pagan ones.

Star Trek comedy:

Kirk: Attention! As you know, we are now on Earth in 1967. Records indicate that by 2290 A.D. this entire planet along with all life shall be wiped out! What do we do?

Spock: I do not know, sir. But it seems fascinating. May we study this Apocalypse some more?

Kirk: Time is running out! We must save the human race!

Spock: You seem to be emotionally hung up on your own kind. Surely we have time. It may take over three centuries for all that to occur.

Kirk: Spock! You are not even human...!

Spock: Kirk, was that a racist comment I just heard?

And so on. You must have gotten the joke by now. Now to our point: Humor aside, we feel that Star Trek was very good in its day, the 1960s, for making some predictions that must be very close to the truth! That show roughly takes place near the Year 2290 A.D. which of course is the beginning - again - of the Age of Aquarius. That is also when Mankind may be tested by some great series of disasters. It is a simple deduction.

Summary:
So let us get into basics: The Bible referred to 410 A.D. in a naive way. Modern science once decided that The End would come by 2000 A.D. Well, now astronomy finally tells us it must be 2290 for real.

End of essay.

Flavian Code: An Omgal Guide to Democracy

Flavius says: "My main concept is simple: Freedom thru trade. Yet there is more to my code of moral values...." He goes on in his speeches to explain the basic concepts of politics, values and Life Itself as seen by his own race. They do not want a State but run companies which do all things for their people. (They still elect an Emperor for some odd reason. That is paradox. But who said aliens were easy to understand?) In any society, there are many functions. Such as religion, economy, education, defense, police, poverty, sex, etc. They have never allowed any one function of any Company - which in our society is the State - to dominate all others. (For instance if the police get too much power then it turns into a "police" state and so on.) This basic concept was always part of Omgal democracy. This was by instinct. They really have very little in common with Eridan. Over the eons, all over this galaxy. Flavius tends to speak for his leader, Emperor Zarcon. Who in turn, evokes The Creator Himself. The basic idea is like Separation of Powers in the Constitution of the United States of America. We can see how Omgalii like Trader Flavius have much in common with American society. Above all, a belief in Free Enterprise. They also can breathe our air. In other words, oxygen. Most of us find his speeches rather pompous.

End of essay.

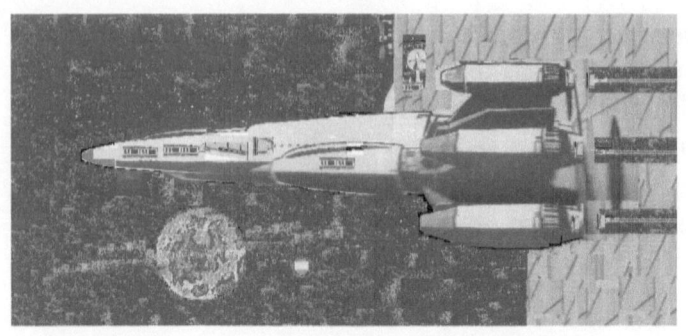

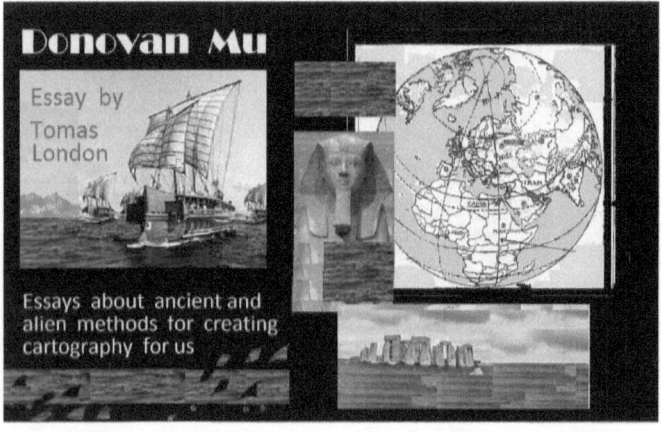

Donovan Mu

Essay by Tomas London

Essays about ancient and alien methods for creating cartography for us

Atlantis,
An ancient Myth Now Demystified

Introduction:
Subtitle: Grid patterns and map based on hypothetical maps from ancient times and also some folklore from the Island Continent of Atlantis. This essay was created in 2005 and was later expanded to become the book Donovan Mu.

Part One: The Prime Meridian

The Prime Meridian on any planet is a fancy name for its Zero Longitude and of course defines any point on its surface as we move from East to West in a horizontal path. Latitude defines points on a vertical path and starts from Zero degrees at the Equator. On our planet, the highest possible Latitude is 90 degrees at the Poles; while the highest Longitude is 180 degrees at the International Dateline.

The Prime Meridian we use was first defined in 1688 by the Royal Astronomy Society of England and has its Zero Longitude on the exac t spot where its first observatory stands in Greenwich near the center of London. This line passes through France, Spain and Western Africa on its way to the South Pole and then turns into the Dateline as it continues up into the Pacific Ocean. That line goes through the Fiji island chain and lies far west of Hawaii. It ends up going through Chukchi Peninsula of Siberia and finally hits North Pole. The Equator is set by natural standards which are based on astronomy. They involve the rotation of our planet along with the seasons. This line was easy to define by observation since ancient times and was useful to navigators. However, there was no logical place on Terra to set the bottom of the vertical sides of the grid pattern for some ancient map, so they used none. Either that or they used whatever was convenient.

Actually, we suspect that nobody ever needed any grids at all until the Renaissance. Roman maps must have had very simple grids, since they only showed Mare Nostrum. Those may have centered on the city of Rome itself. Later on they may have used Constantinople, which is almost in line with Alexandria. My conclusion: the current line at Greenwich is arbitrary and was invented only recently in the course of History itself.

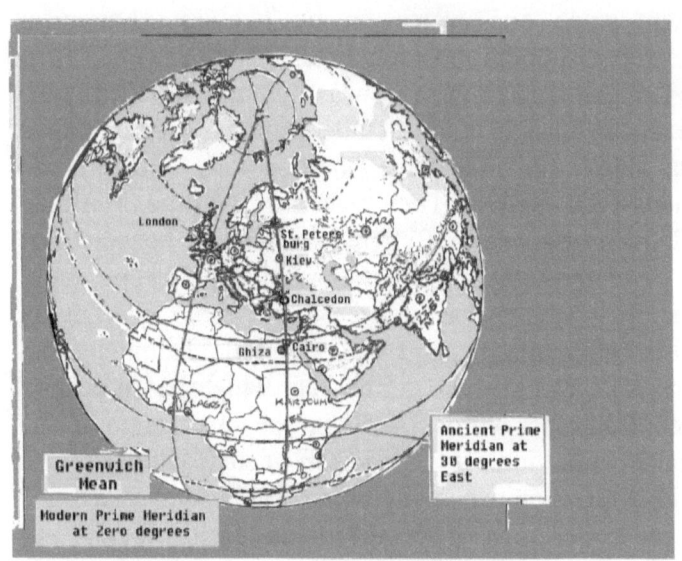

The Ancient Line at Cairo:

My own theory is that long before our ancient empires, namely Persian, Greek and Roman, we had a certain prime meridian of Longitude already. It was probably not invented by the Sumerians themselves, since their astronomy was primitive and suited for practical uses only. They could not have known that the planet was a globe, and the fact that it was exactly 7926 miles in diameter. Any grid was arbitrary but found later that it was best to base it on twelves and not the decimal system, since our cycles go by twelves. (Each year has twelve months and each day has 24 hours.)

The grid I have in mind was invented by aliens who had the means to make detailed surveys of our planet from above and may have done so long before homo sapiens even existed. They placed their Prime Meridian at Cairo (or the nearby Pyramids of Giza). The exact line must run through the Great Pyramid of Cheops itself, since they say it was faced with durable and highly visible white marble and had a tip of gleaming gold. That alone would indicate markers which had precision in mind. In other words, I feel that the very design of the Great Pyramid indicated some sort of survey tool.

One question can easily be answered: why place this marker in the area it is in? It is easily found and highly visible. The Nile River flows for hundreds of miles in a nearly straight line along the Red Sea and hits the Mediterranean at right angles. The surrounding land is yellow sand so the river and its large delta (which are dark) will contrast with it sharply. The clear desert sky makes it visible at all times, even at night. Needless to say, this part of the world never has ice cover not dense foliage. It is on a plateau high enough not to be flooded or silted up. Nor do any nearby mountains clutter it up.

There is a better reason for this location. Cairo is right at the geographical center of our planet's major land masses, which can be confirmed by taking the Dateline as edges of your global map and then placing Africa and Europe at the rough middle horizontally. Keep in mind that the Pacific Ocean is gigantic and that the Old and New Worlds were once one continent and are still very close.

The famous Turkish maps (found in 1513 by Piri Reis) are said to center on Antarctica, which is taken to be proof of photo surveys from high orbit. Well, my essay has nothing to do with that concept. Those maps would not be useful l to any ancient human being, and will not be relevant to any aliens, unless they wish to set up bases on our coldest island continent.

[**Editor:** How scary. So you must have been influenced by Jerry Pournelle's **Tran** books. Must have been influenced by von Daniken, who may have been inspired by some obscure book from 1965 by Arthur C. Clarke. Am I correct?]

Again, that Meridian at Cairo:

My ancient "zero" line lies exactly 31 degrees 12 minutes East of our modern line at Greenwich and 30 degrees North of the Equator. This line cuts north close to Istanbul and through Kiev and Saint Petersburg. But more pointedly it will turn into a New

Dateline that bisects the Pacific but 31 degrees East of our own (in other words at about 150 degrees West as we see it). That cuts through only two major places: Tahiti in the South Pacific and Anchorage in Alaska. (Also Mount McKinley in the same state.) My new dateline is now East of Hawaii. The Zero Meridian in Africa now goes past close to Khartoum, through the ruins of Napata, the Sudd marshes and then Rwanda. It cuts thru the Rift Valley to emerge from Durban. Maybe we should call this line across the very peak of Cheops, Ancient Meridian, not new. As we know, our modern version was created in 1688.

Stonehenge as major factor:

Stonehenge in England is close to London. In fact it is two degrees West of Greenwich. That makes it 33.3 degrees west of our ancient Meridian at Cairo (or Gizha but that is getting picky) which is very important since it reminds us of the famous "33" of the Masonic movement. Could it be that their main mystical number came from some ancient map based on monuments in Egypt? This third line goes thru the extreme Eastern part of Maroc at the town of Oudja, cutting exactly thru Al Jebal which is listed as a "fortress" on a high plateau. Among the Atlas Mountains. Then comes out of Africa near Accra in Ghana. The sands of the Sahara may cover many copies of the English Stonehenge. We think it was built in 2000 BC by some Phoenicians who had actually annexed part of England and who used metal tools.

Section Two:
Some speculation on construction method:

We have several theories about the Pyramid Complex of Ghiza at Cairo and they all point towards the conclusion that this site was planned as marker for some ancient map. It was also a monument dedicated to astronomy, alien or not.

Why so? Well, in the first place why mention aliens at all? It is possible that the human race is from this planet so that all ancient structures we have on Terra were built exclusively by Cro Magnon during the past 12,000 years. To do so would require metal tools because stone tools were too slow and primitive. The large public buildings of Egypt (mostly devoted to religion) were assumed to be built between 3000 and

1000 BC. They are of such good quality that an advanced system of engineering was used. The quality of the masonry alone is such that steel tools must have been used. Yet, according to modern History the earliest iron was smelted in the Mideast circa 1,200 BC. (We admit that some brass and obsidian is hard enough for mason work, even bronze saws with diamond edges, say. Even by 5,000 B.C engineering skills were amazing.) Again, not all knowledge has to come out of flying saucers. This is a moderate, flexible point of view. It is very popular.

We were tempted to disagree with that common theory for now by personally assuming that iron and steel were smelted long before the usual date, but that must be extreme. A pointless controversy. There are some reasons for this: Iron is of great strategic value when tools are of mostly of other metals. Copper is softest. Then comes bronze. Brass is hardest of the non ferric metals. The problem is that to create iron, blast furnaces must be used to get pure metal from
raw ores due it's the much higher melting point. These furnaces use fires that are far hotter than anything used for the other ores. They involve coal of some kind and large towers which funnel heat into a small area. Since ferrous ores often contain impurities such as sulfur, they have to use chalk powder to absorb and slag off these elements. That makes it a relatively complex refining process. This can only be done in areas where three raw materials are found: coal, limestone and iron ore itself. Instead of coal one may use wood as charcoal but that is wimpy.

We might assume, in this bogus theory, that the famous Hallstadt Culture (villagers who made bronze in Austria in 4,500 BC) was only an atavistic leftover in the hills. Meanwhile in the valleys more advanced urban centers had iron technology and were already planning an invasion. The Nile, Indus and of course Tigris Euphrates valleys had iron at that time and it may just be that Rhine, Seine, Thames and Danube valleys may have had similar abilities. But why bother? I'ts a long shot.

What we are doing here is offering a strong rebuttal to the old theory of how us natives of Terra were too dumb to invent anything and so we had to depend on some aliens who somehow landed here and decided to help us. That idea comes from Eric von Daniken, an author who was and is, legit, but who must have been inspired by Clarke and **seems** to think that we humans got all of our culture from aliens. [**Note:** Some people seem to think of him as a con artist, but we disagree. He has been doing his work faithfully for decades. As in, going into jungles and deserts and measuring and photographing many ancient monuments. New editions of **Chariots** now have color photos and far more data. He also, like many academics, may not speak English well. So his image suffers. He also tends to be objective, like any formal academic.]

More or less, we can call the following a "leftist" view: As in, we do not assume everything we poor beings know & have as Earth Culture comes from some alien freak. OK? In general, in our modern society, those who feel that the Human Race has evolved here on its own according to Darwin's and other secular theories are often called **The Left,** or Politically Correct if you wish. Or P.C. And conversely, those who feel that all of our genes, science, religion, and yes, even maps come from aliens and/or gods are usually called The Right. Or P.I. And so forth. **Our PC and PI?** An unholy mess. Okay?

Death from Above :

The Sumerian texts have long passages dealing with the waging of war between city states and empires, almost as if that was their only pursuit. Anyway, one book constantly repeats the phrase "let not one stone remain upon the other" and "rend their cities down to the very foundations". One text has these two phrases repeated at least ten times on one page. The Bible quotes these ideas as well.

Why such mindless hate for defeated enemies? I feel that the real target in these nasty books were the blast furnaces found in major cities of that era. Since steel gave cities strategic advantages over each other it was advisable to destroy the industrial base of some defeated enemy. Iron must have been of great value in terms of money in those days, far more than copper, silver or even gold. It was hard to make, rusted easily and vital to industry.

This is one reason for the lack of evidence of any steel industry before 1,200 BC. The stuff rusted away and any scrap left was stolen and melted down long before we found any ruins. Stonehenge stands in a humid land. Yet iron rusts even in the arid climate of Egypt since there is condensation overnight and water just under the surface. Any iron tools left behind by any pyramid builders was taken by various thieves long before Islam arrived. Actually it is even possible that copper smelting existed by 8,000 BC and that iron soon followed.

Our joke theory about iron has one flaw: why was all this not mentioned in records before 1,200 BC? Well, the languages we now know were not the same as Sumerian. We can figure out many terms in relatively ancient languages like Latin, Greek, Phoenecan, Persian and Hebrew. Yet we have no exact words for such metals in older tongues. Basic Sumerian comes from long before 5000 BC. Later dialects such as Akkad of the Chaldeans started circa 3500 BC. In all of them the only word for "copper" is said to be "ur" and that is rather silly. Maybe "ur" was the root or our word "ore" and that can refer to any metal element.

The Sumerian language has to be more specific than that, for there were at least six base metals (copper, tin, zinc, lead, silver and gold) smelted plus some alloys. Each was a distinct element so that the word "ore" (referring to various oxides) must be generic. Thus we feel that the old word "Ur" had many meanings, such as just basic junk containing some element, or maybe only their most basic kind utilized, which was copper itself, the most important commodity for trade, and finally the name of a famous Chaldean city. This is what millennia of linguistic drift lead to.

In other words, all metals came from ores and the ancient Sumerian word begat our modern word. They sound alike. The old words for our six ancient metals were probably something like "red ore"..."white ore"... "yellow ore" and so forth. As in, Red for copper, Yellow for gold, etc. The words used so long ago in Mideast lands were enigmatic. Very abbreviated.

But my real point is that the technology behind iron was kept as a military secret in ancient times until certain more recent empires took over. Even they must have censored records of iron until finally their elite became more confidant and figured any further secrecy was pointless. One more point can be made: Gold was usually not used in for currency in the New World. It became so in the Old. (Yet that is not relevant to this essay.) Basically, any strong nation with some kind of steel industry tried to keep others from having it also. Why not call it "bombing" some enemy into the Stone Age. Or Bronze? Which is not funny. And that ends Section Two which deals only with construction methods involved.

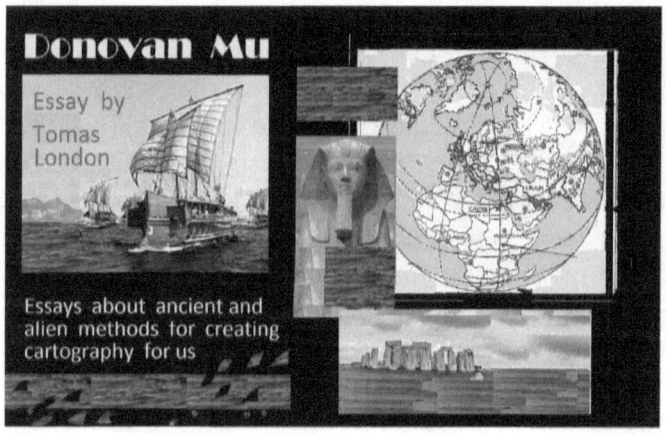

Section Three:
The Function and End Result of Pyramids

So we have this amazing group of pyramids at the geographic center of the world's land masses and it has maximum visibility. But what was their function? That basic function must have been as Prime meridian (or baseline) for some ancient grid. Thereafter it was easy for ancient surveyors, astronomers and sailors to check out the facts themselves and make maps of their own.

The first pyramid might have been built long before the usual date of 2740 BC. That idea comes from most Atlantis dreamers. Being inclined towards the mystic, they tend to claim that people from Atlantis built Cheops long ago, even before the last Ice Age. Which must make it over 25,000 years old. In which case, only aliens could have done it. But all of this is speculation.

What we can say, however, is that the most important date recorded by this complex is 19,310 BC because that was a vital date in basic astronomy. It is when the change from one galactic "eon" to another occurs at the Vernal Equinox. This period contains twelve Signs of the Zodiac, each of which last 2,160 years. It ends in 6,610 A.D. and thus lasts 25,920 years. This is also called a Platonic Year. Our planet tilts away by 23.5 degrees from the baseline of the Solar System and also the galactic equator. This has an effect on climate as well and a lot do with ice ages. The basic point here is that every 25,920 years another Eon begins. I am using the word Eon for the sake of clarity and keep in mind that the word Age is often used for Signs of the Zodiac. Such as the Aquarian Era which begins in AD 2290.

The reason we bring up Platonic Years is that the Great Pyramid can have a function related to this Platonic Eon. It involves the two stars Vega and Polaris. Every 25,920 years the stars switch roles as the True North on your compass. One shaft inside Cheops is only a few feet wide and so will focus exactly on only one star. This shaft runs inside for hundreds of feet and forms a perfect telescope. It is on the north face of Cheops and thus will focus on the two polar stars. To confirm my theory, the exact angle of the shaft as it relates to the pyramid's position on Earth must be measured. As I say it will prove to be relevant our two northern stars, which makes Cheops into an observatory. It goes without saying that such a durable monument as this was useful as an instrument for determining the accuracy of any further tools our "race" needed for better calendars, maps, navigation tools and so forth. The accuracy of the grid itself could be checked out from time to time over the ages by this method. Thus Cheops was primarily a grid marker and secondly an observatory.

Unless the data itself comes from aliens the Sumerians provided the original geophysical data for

Cheops itself and the rest of the complex was added by the Pharaohs, who came later. It may have had dead royalty walled up inside for mystical reasons: to sanctify it. That was just for the sake of ritual, since I feel that the original builders were the smartest. We suspect that killing someone and then sticking his dead body into expensive public monuments was often done in those times. In any case that creepy act must have inspired the expensive funeral customs of Egypt.

Why did we say the above? Because the smaller pyramids were not as well built and they get progressively lousier. They get smaller and sloppier and there is no coherent plan as in other ceremonial centers like Teotihuacan or any Mayan city. My argument is that the original purpose of Cheops was not as cemetery but for the more secular one I have deduced. Here are some details: as the dynasties of Egypt went on they got more decadent. Egypt had advanced surgery but there were some problems already. Probably only the rich could afford it. Not slaves. Yet those very same elites wasted the medical talent of their nation on bodies that were already dead. It was also a waste of money.

This society was openly operating in a manner that was not as rational as others. They may have forgotten the true function of Cheops long before it fell to the Greeks in 332 BC. The most important fact about this monument is that it expressed astronomy on a modern level. The grid it supports and the results of it being surveyed and used, eventually led our ancestors to conclude that our planet was a globe. Which could be divided into like, 360 degrees vertically. This led to such time periods as the Sumerian Sar of 3,600 years and the Mayan Baktun of 360 years. Mayan calendar also has 36 months of 10 days each, indicating a year of 360 days with five days extra for some evil festival, but more to the point it does conforms to our grid.

In a later section I will demonstrate that links between Sumeria and Central America may have existed as early as 3000 BC. The actual travel will have been done by later cultures such as Phoenicians, also Egyptians and Greeks. All of them were good sailors and capable of exploring the oceans at will. But what is the point of this section? To show that no matter who first designed this grid and its giant marker also made the entire ancient world believe in our planet as a globe revolving around its sun...which we now think also but with a new baseline in England and only since 1688.

About Central America

The Mayan Connection:
This part deals with possible travel from the Old World to the New meaning before the Vikings who existed after AD 800, and we mention Columbus as well. The whole idea behind this essay was to show that there must have been maps older than the ones we have been making in our modern society for the last 314 years.

Greenwich Mean was set in 1688. We are implying from the start that our oldest Earthian grid was designed by aliens who came from another star system, and that surveying was done from above long before They bothered to land here. Since we can do the same sort of surveys and even make maps of other planets by using probes, any such advanced Beings will have none.

My ancient grid based on the Longitude of Cairo must be older than 2740 BC because all major nations in Europe, North Africa and the Mideast from then until the Fall of Rome behaved as if it were valid. Various nations of what we call Antiquity achieved brilliant advances in navigation for until the Church of Rome finally denounced this in AD 410. We are not saying that aliens had to be the sole inventors of this grid because Sumerians could have easily designed it themselves; later nations may have likewise built monuments at Cairo for the sake of

Science as it was then known. Finally the Phoenicians of the Levant, who were known as masters of navigation, tested this grid by sailing all the way to India, Greenland and perhaps even around the entire continent of Africa. It is well known that the three major nations I am studying come in this order: Sumer, Egypt and then the Levant. They played their roles well and may have done so without any flying saucers.

So what about the Mayans? Well, it is said that in 3114 BC their culture began in Central America. That was also at the height of Chaldean culture, which is more recent and more accessible to us. Since there exist legends of whole armies trekking off to Central Africa to make war (these came from Sumerian records) it may be that they actually sailed all the way down the Nile to its source in the Congo Basin. The same people may also have sailed down the east coast of Africa as far as Dar As Salam. They may even have gone West of Egypt to pass through the rocky

Gates of Hercules at Gibraltar right down to Senegal. It was possible for Sumeria, but easier and more likely for more recent cultures like Egypt and the Levant.

[Pause for an earned breath by any tired academic. For you see, this be 100% academic. Leaving out the usual braggadocio and lousy humor as found in our fiction, We are Brethren. Then come aliens. That means Things that really cannot be one of us.]

The logistics of such travel are easy to figure. We all know about reed and balsa rafts from the old Kon Tiki movies. However such craft are not important to this essay. None but bandits or refugee will use such small craft. The risk is too high. Long before the Romans there were large galleys that crossed Mare Nostrum with ease. The biggest were not even warships called Triremes. Those had both sails and oars, which made them better some later ships. Even Egypt had barges and cargo vessels over 200 feet long. Here are some stats: the distance from Senegal (the western tip of Africa) to the eastern tip of Brasil near the city of Recife is only 1800 miles. The distance that Columbus crossed in 1492 in three small ships was 2.4 times as long as that which any ancient Old World craft ventured into when voyaging from the Old World to the New. They kept close to land until they hit Senegal, then headed West.

That was an amazing feat because the trip from Spain to the Caribbean island he landed on was 4,200 miles. The water that lies between Senegal and Bahia is often calm and has a better climate. Under such windless conditions some trireme with many rowing, can cover 1800 miles in less than one month. We have no idea why Columbus crossed the Atlantic at its widest part. He must have known that the Arctic route was shorter but he was actually looking for Indonesia which is tropical. He was certainly not looking for Atlantis because as far as he knew, it did not exist. At this very point Columbus becomes relevant: If our grid was any factor then it may have made Columbus think that Indonesia (his target) was about thirty degrees to the east of where it really is. Has that point ever been considered by historians?

A word on Byzantium: It fell in 1453 which means that during the next four decades many refugees came to Italy and Spain to survive as Christians. They must have saved records from the Turks. Among them were maps which may have come from ancient sources, and so were used by Western navigators. At the same time Vasco de Gama and Amerigo Vespucci had similar ideas but went by other routes. The famous land routes, such as Silk Road over Central Asia and caravan treks over the Sahara, were far too costly for rational trade and only existed during the Dark Ages because water travel was blocked by constant warfare. The Byzantine and Islamic empires were despotic to the extent that they censored all intellectual activities. Including funny maps from antiquity.

So finally, you may ask, how did that first expedition from our Old World make it to the New World? Simply by following two routes: The southern by leaving Mare Nostrum through the Gates of Hercules. Then follow the coast of Africa until their fleet hit Senegal. Then a short hop of 1,800 miles across salt water until the extreme Eastern coast of Brasil looms up. Bang. That's Mu right there. The second route was by heading North past Gaul by ship, passing Hibernia, which we now call Scotland, and following an eternal frozen shelf of ice (just like what surrounds Antarctica) until they hit Vinland, which is the Viking name for the Eastern tip of Atlantis. Get it? Both routes are easy to flollw, and both offer food & fresh water. We think many such expeditions did so during the Bronze Age.

In fact, there is a theory held by academics in the University of Rio that, in 1753, a certain Bandierante called Joao da Silva G. wrote of a "lost" city somewhere in the State of Bahia. He said that it was built by Chaldeans, which means around the time the pyramid of Cheops was built. Rings bells, right? Like the Portuguese, they were much earlier survey missions trying to chart our globe by sailing to Mu. Another lost continent. So it is clear by now that this myth started as fact about the same time as the above voyage, namely in 3114 BC. That date was also the official beginning of the Mayan calendar.

We know culture moved West from Sumeria from 5,000 BC to the time of Plato, 360 BC. The Phoenecians founded Kart Hadasht (modern Carthage) later as their bigger and better financial center but must have explored and even colonized other parts of Mare Nostrum first. They mixed with the Celto-Iberians of Spain plus the Berbers and Tuareg of Maroc. Then they sailed up to England, eventually to build Stonehenge in 2000 BC. They continued into Ireland, Iceland and finally Greenland. It has been recorded that many European (meaning white) tribes travelled and traded on a regular basis across the East Atlantic Ocean and must have had contact with Canadian natives and Eskimos. It has been said by some that Vikings made it to New Brunswick, which they called Vinland. They may even have ventured into the Saint Lawrence River to Montreal. The real name for this area was of course Atlantis and it was called an "island continent" for good reason. At that time it must have seemed to be the same size as Ultima Thule so the older names were quite suitable. But actually, had they known the full extent of North America they still would have applied the same term to it: Atlantis of Yore! Monty Python: "Hey. It's only a model."

The Old World is so compact and extensive that it contains most of our land surface. Yet North and South America are small and isolated so that both can be called "islands" but still big enough to count as continents along with Ultima Thule and Australia which may have been Mu/Lemuria). In fact, the British Isles have been classified by the Romans as an isolated continent. Pause for effect.... They called it Hyperborea. (See more detail below.)

So there we have it. Even our term "Atlantic" itself must have meant "global" since this must have referred to that entire expanse of water between Japan and Spain. To ancient scholars this was simply one big World Ocean... with an island called Atlantis in the middle, and South America as Mu? Why not? Keep in mind that the Pacific Ocean was not called that until about 1700.

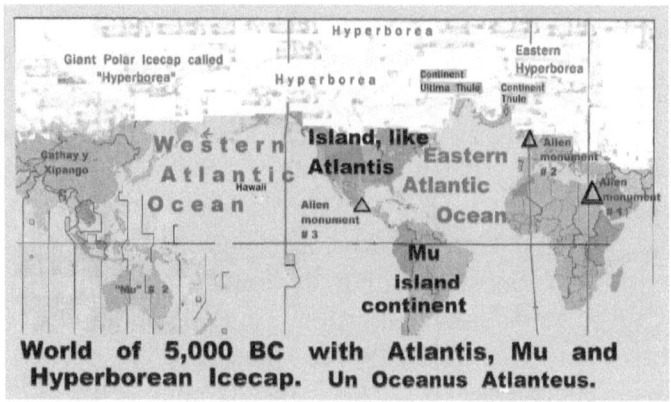

World of 5,000 BC with Atlantis, Mu and Hyperborean Icecap. Un Oceanus Atlanteus.

* Special note on Hyperborea:

It is obvious to us that our planet's natural Ice Ages had a lot to do with these polar land masses. That, of course, is based on our basic cycle of 25,920 years. Aside from any industrial pollution, greenhouse gases, and the like. In very ancient times, one giant mass of ice covered the entire Arctic Ocean. This of course hid the fact that Greenland, Iceland and even British Isles were all connected by ice. (Right now southern England is subtropical, but that was not possible back before the Birth of Christ.) That mass extended to Scandia. Stone & Bronze Age hunters could easily have wandered all over this like Eskimos. They had plenty of fish, whales and seals to eat. They did not even need boats. As for fresh water, it was there as snow. But any evidence of this would now be lost under miles of water.

Eventually, this ice shelf melted away. By the

time of Vikings, the British Isles and Iceland were left over as "island" continents. Only the southern tip of what we now call Greenland was bare of ice, and had some grass and stunted trees on it, which allowed a few natives called Skraeling and Europeans to live. But what is our point? Very simple: The first Greeks called all polar regions by the name of Hyperborea, meaning the Land of Boreas, god of wind & cold. Later on, after some expeditions to Atlantis (as in Quebec & the Maritimes.) They named these separate land masses Thule, Ultima Thule and Vinland. Ireland was called Erin, and Scotland Hybernia. Only England itself kept the name Hyperborea, which had become an official name on Roman maps. Each of these places was of course considered an "island" continent. They did not realize that Atlantis & Mu were far larger than any of these Boreal lands combined. They were, however, convenient stepping stones across their East Atlantic Ocean. This entire argument is based on shifts of logic and semantics. Or should that be linguistic drift?

The above note was written in 2017, long after Tomas created his original Atlantis essay in 2005. He was inspired by a certain Disney movie of the same name, which is worth seeing even for adults. Later on we will bring up this issue again, the word Hyperborea being a key word. Try to ignore our humor. A word on Mu according to Churchward: He made a major error by saying it was in the Pacific and also that it "sank" really deep under the waves. You are invited to google Mu and then to explore some of our familiar Mid Oceanic mountain chains to grok this.

Plato and his Atlantis Myth:

So what became of Atlantis? Plato said it sank under the ocean after massive earthquakes and that thereafter the area was haunted by demons. Both ideas seem absurd today. We feel that Plato had very good motives for such extreme claims. He made them in 360 BC, which came just before the rise of Alexander the Great. We therefore suspect that Plato was a Greek patriot who was paid to make up propaganda as it suited the elite of his nation.

Humor aside (there are many jokes about Alexander) we should realize that Greek city states back then covered a huge area, which included Italy, Sicily, Turkey and the Balkans but was not unified. However, this ethnic group was on its way up and ready to become the new "superpower"... next to Persia, who had always been the eternal rival of any major empire. The Persians were safe up on their plateau and hidden behind deserts and mountains. Before the sudden rise of Macedonian power (which was soon to lead the rest of the Hellenic world) Persia also had the strongest army of the world.

Macedonia had for long planned to destroy Persia totally, but its leaders realized that superior land forces were needed. So funds had to be diverted from the building and arming of ships, since the Hellenic world was the major naval power of its era. As a result, about 340 BC the Macedonian elite began mobilizing their giant war machine, but paid only for infantry this time. The idea of Greeks not having the world's best navy (aside from Carthage and Rome?) was shocking and they needed to rationalize it. Hence this freaky myth about large land masses getting dunked. For the next 1.200 years nobody sailed between the Old and New Worlds.

The distance between Arctic islands was small but conditions up there were lethal. The Vikings made some progress between AD 800 and 1200 but finally gave up. They were too primitive to survive. More relevant factoids: These people were uaually illiterate and only spoke Germanic dialect. Few knew any Greek or Latin. Also, they had no real maps nor proper instruments like astrolabes. But at least Vikings were smart enough to practice island hopping a term used by USAF pilots in World War II. Even for Columbus, it was a challenge to cross the North Atlantic where he did, for his largest ship - Santa Maria - was only 80 feet long. That was much smaller than an ancient galley, which were at least 125 feet long. Those things were less likely to roll over in high waves than medieval caravels and did not depend on wind to move.

To be honest, why did Columbus not do what the Vikings did? We still need to analyze his logic. It was not even based on the survival instincts Vikings had. Why did he have to sail at about Latitude 30 North? Because he was motivated by Pyramid Logic! He just went by what ancient charts based on Cheops told him, and nothing else. Such as, if at the Latitude of Cheops, namely 30 North, the circumference of Earth is exactly 21,600 miles, then that is what he decided was "right" for his own personal plans. Again, his personal concept of Earth As Globe was based on charts ancient Sumerians designed.

Note on Columbus:

Here is a vital ethnic note: Columbus was from some kind of medieval culture, but not Germanic. Nor Celtic. His mind was based on Latin and Greek concepts only. And certainly not anything before
that, such as Levantine, Egyptian or Sumerian! Why? The Rosetta Stone had not been found until 1800, so that none of those ancient sources were available to him. No humans in his time understood those languages. As for Hebrew or Arabic, what? His society totally rejected both Arabs and Jews to the death. By the way, as for native Atlanteans, here is my theory based on basic logic: Racism. He first landed on Cuba. Then he hit Hispaniola and made solid claims for Spain. By the time had seen his second island, he must have decided that his naive theory about that region was wrong! Many of us have

suspected this. Yet records tell us that, until Cortes came along decades later, most Spaniards still actually believed they had landed somewhere in tropical Asia. Why not? Indonesia, Cambodia and India may have had pagan societies similar to those of Latin America. But what about China and Japan?

Marco Polo had started a subtle but strong tradition of respect for Buddhism and some Islam. These cultures were at least accepted by intellectuals like himself. Monotheism was okay. But here, the Indios practiced snake worship? Human sacrifice? That was too much. To any Spaniard, these Mayan and Aztec cultures had to be extreme forms of Satanism. Clerics must have assumed the Mayans were then followers of the Hindu goddess Kali, which also featured skulls, snakes and identical junk. The Veil of Maya. That led to the name. Los Indios de Maya, This cult had to be destroyed. War on Natives began before his return voyage, which was of course by the ancient route. Now we can see the logic Columbus must have used. These people were real scientists. The Dark Ages ended by 1492.

NOTE: [This is nerdy and may insult your I.Q. level, but why did our Natives not have gunpowder and steel? If they had, then they might have resisted the Conquistadores. This question will be answered in my next essay, Fuck Atlantis Redux. It will be short and sweet.]

This is an essay about cartography. Not flying saucers. Our central concept is based on Ontology. Human logic and how it can change over time. Our concept of "island" continents has to be strictly an ancient one. Not modern. It has no meaning to us, when we can just travel, trade and fight with modern methods. Planes, rockets and the Internet. Take one Roman concept as outdated: Hyperborea. Again, until Caesar invaded this "remote" isolated "continent" that was the common name for all British Isles. In Latin of course. And Greek before that. No kidding.

We are now being really pushy. But get this: The Stone Age word for "water" and root for many modern words such as "atavism" is "ata" so why not consider that "Ataland" was the very first word for North America. It was first discovered 12,000 years ago by some tribes from Siberia. That Stone Age name soon changed to... Atland? Later in Greek to... Atlantis, similar to Colchis or maybe Sardis? The Greek Empire claimed in their hubris to conquer the whole world, so why not Hellenize their maps? Now why did Plato avoid simple facts like that? Because he made many things up, we think.

We do have access to people who took Linguistics in University. It also helps to surf the Internet. As for our World Ocean Theory: Ata is water, which covers most of our planet. Ergo, Atlas as a person or god can still carry the world on his back. (Why is Atlas a man? They just needed the concept of men as symbols of power. Same as Hercules. A man can hold up a whole planet. Women are too weak for that eternal task.

Uncool, eh?) They even used to think that all land floated on water, which was an Abyss. Bottomless. (The Sumerian word was Absu.) That is correct in a way, since our crust floats on liquid lava. (Yet one more example of Logic from ancient minds which to us seems naive.) Hence all salt water on Terra was one big ocean called The Atlantean.

Now we know what that old Atlantis myth is all about. The land mass itself was real, if taken to mean North America but its factual database was converted into fiction for military reasons. In other words, the economy at some point forbade any idealistic voyages across what was then considered to be the "Eastern" half of the "Atlantean" or, World Ocean. It makes sense. We have left out some mythology that deals with Leprechaun and Alien events both in legend and modern experience. These freaky myths may be the real cause of my grid but will not be accepted by academics. We have also repeated the same points like some blowhard. Forgive me.

[That stuff shall appear in the next essay.]

Notes on History

Reality in detail on Cheops:

The First Dynasty started in 2775 BC with King Snefru but the next monarch of Egypt was the actual builder of the Great Pyramid himself. That was circa 2700 BC (his name was Khufu in his native tongue and Cheops in Greek). It is hard to tell the actual purpose of his project from the facts because they are vague. In the first place they were just at the beginning of recording history and nothing was stated directly about funeral customs as they related to the pyramid itself. It was said that while Snefru was a "good" king, Khufru aka Cheops was "evil" or tyrant but that opinion came from irate taxpayers which is normal. In truth we don't know what political ideas these rulers had so the purpose of their architecture is unknown. The facts are these:

Before Cheops it was custom to devote land next major pyramids to some kind of necropolis... which they needed anyway. What is relevant to me and Galaxa on a personal basis is this:

Cheops rested on solid sandstone. That was its "load" bearing part. The "jacket" in our terms was made of thin sheets of white marble. The very tip was some shiny yellow metal like brass, which was said to be gold. (Do you believe that?) Some of that marble is still up there near the summit, but most of it was stolen long before anyone Islam appeared. Marble is valuable. It must still be around in the form of nice palaces for some Caliph. Dumb they are not. As a joke, Galaxa must think Her slabs are now "cladding" the tallest office tower in Canada. No wonder She and otros Leprechaun folk like to dance about in First Canadian Place in short skirts. Twas only a lark so they vanish in a moment. Why not? [Back to reality.]

Later on, as grave robbing became common, most monarchs chose to be buried in graves that were totally hidden, far away from any open monuments. They may have even staged mock funerals within major pyramid complexes to distract attention from real tombs. To summarize, we feel that all those elaborate funeral customs of Pharaoh were designed as a cover from the beginning. To hide secular activities. Meaning Science, of course. Final note: Both towers have a high albedo, thus are very visible among darker objects. They are nice tools for astronomy.

[Note: The above is symbolic logic. Try to think like an alien. Like forget your own concepts. We call this Exobiology.]

The Scholars of Alexandria:
Soon after conquest in 332 BC the Greeks founded this large urban center. It had the world's biggest library until AD 400 and aside from any alien influences they had whole communes of academics who worked on advances in science. They were close to reaching our own modern standards in the Third Century BC. Of these it seems that Eratosthenes was the best ancient scholar.

Eratosthenes lived from 275 to 194 BC and in 240 BC wrote his book "Geographica" which dealt with the "spherical nature" of Terra. In it he calculated the circumference of our planet based on simple land surveys done locally. He also tried to design a global grid based on those concepts. It was our first geodesic team project and we feel that most of it came from Cro Magnon minds.

This data remained in circulation until the Fall of Rome. There after the Western Church banned the above book and similar data and much of this was actually destroyed in the anarchy which followed. The facts so far are famous but some of this "weird" data must have been moved to Byzantium and stored secretly, probably under the Hagia Sophia itself. This latter bit of file saving from some ancient data panic Roman Style is not as well known because the Byzantine Empire and what it stood for have always been elusive. A dark part of Western History.

Final Note on Cheops

This comes from the book "In search of ancient astronauts" by E.C. Krupp published 1977 by Doubleday in NYC. In the diagram from that book are shown the shafts: the southern one points at Belt of Orion while one of the two on its north face point to the star Thuban at 31 degrees. The lower one points up at some other star but this is not named. According to astronomer Chris Dolan, the star Thuban was Polestar in 2990 BC, meaning that back then, it held the place that Polaris is in today, which is always directly over the North Pole. In AD 2290 Thuban will return to that position again. The star Vega also changes places with Polaris in this way but that will happen 14,000 years from now. The shafts of Cheops seem to be pointed at prominent stars and so must have much to do with this Vernal Procession we have mentioned so often, yet the author being am quoting here has not elaborated much.

My conclusion is that the northern shafts are designed to show major stars as they cycle through the polestar position. Since they knew when that was to happen, this whole thing was and is useful as a calibration device for calendars and maps through the ages. Ancient humans knew about galactic shift and accounted for it. In fact, we even suspect that ice ages came and went according to our familiar cycle of Eons of 25,920 years. This opinion is based on books having a technical nature and not mystical stuff from the 1970s. All of them are easy to find.
Main essay ends here.

Part Two: Mini essays on Terran Cartography

Note: this essay has many parts to it. The first is one on Mayan system of math and calendars. Then come two pages on some problems we had with Baktuns. Then come two pages about Stonehenge.

New notes on Mayan System:

They had a basic system of 36 months of 10 days each. That gives exactly 360 days and they always hid their five "nameless" days. That may have been a period of five days, from July 17 to July 23. That would have been filled with drug parties, pagan ceremonies and even human sacrifice.

They had two calendars. The Short and Long Counts. We think the Short version was based on Permutations Of Twelve and based on Sumerian astronomy. The Long was based on our Decimal System and was a

Mayan innovation. Within these two calendars, we can find the numbers "30" and "33" which alludes to our two meridians at Greenwich & Stonehenge. We just took 360 and divided that by 12 which gave us 30. Then 400 by 12 which gave us 33.3333 into infinity. The point is to accept that the two Mayan calendars are based on the difference between two major Math systems: Decimal which evolved from us having ten fingers, then Duodecimal, and that evolved from the rhythms of our Inner Solar System. You can check it out.

KIN= modern day
UINAL=month of 10 days
TUN=year of 360 days
KATUN=20 years
BAKTUN=360 years
PICTUN=7,200 years
CALBTUN=144,000 years
KINXILTUN=2,880,000 years

We are assuming a Sar of 3,600 years and Baktun of 360 with Tonal of 72 years, or one lifetime. This way is more accurate in astronomy and also more in tune with other cultures. Here is the Long Count, or Decimal version:

KIN= modern day
UINAL= 20 kin
TUN= 18 uinal/360 kin
KATUN= 20 tun
BAKTUN= 400 tun

PICTUN= 8000 tun
CALBTUN=160000 tun
KINXILTUN=3200000 tun

Modern Astronomy comments on Mayan:

As you can see, we are trying to prove that Mayan calendars are based on earlier Sumerian astronomy. In other words, this data was taken to Central America for testing in new sites before 3114 BC. It is possible that none of those expeditions sent any results back home. Anyway, ancient Mayan and Hindu calendars are similar. Relevant facts are: Earth has diameter of 7,926 miles on Equator. Circumference: 24,901 mile. The same at Latitude 30 is 21,600 miles, with a difference of 3,301. This was calculated by knowing that while at the Cairo Latitude, we have 60 miles per degree, we have about 70 miles at the Equator.

If we divided the **Platonic Eon,** as in the famous 25,920, by 360 we get exactly 72. The diff between the Eon figure and the real belt at Equator is 1019 miles. We can get more obscure stuff going: On surface we can divide 24,901 by 12 to get 2075.083 so that would have been the length of one Zodiac Era instead of the usual 2,160 years. On the Latitude of Cairo we get this result: each Era Zone will be 1800 miles or if we had divided this belt of 21600 miles into ten then we will get twelve zones of 2160 miles each. Hence the tendency of having funny years of ten months each or even as some Mayan have 36 ten day periods per annum. They tried to compensate for curvature of our globe even back then. There are many ways of expressing the latitude of 30 North in figures that mix up time and space yet still involve that great Cheops complex. That is where most ancient cultures reside. The Eon figure is based on the same Hex System that the ground figures are, only that it will have to be in orbit over planet: We have a rough guess of 150 miles above the surface at the Equator. That will be nice for survey craft.

We can also use simpler method at figuring out where our Mayan 72 and 52 came from: just use compass degrees. Divide 360 by 5 to get 72. Also: 7 x 52 = 360. We now have reason for Mayan system of 52 and 72 steps on their pyramids. The whole thing involves "sin" and "cosin" of our round planet and latitude as compared to orbit as well as length of days. Notice above that a series of Fives and Sevens is involved. Well, that just reflects another major fact about our Solar System which is this: Most motion inside the Snow Belt (same as that well known asteroid belt between Mars and Jupiter) goes by Twelves; but motion outside that goes by Fives and Sevens. Newton much later on, after they finally invented telescopes, had to invent calculus so that we could understand and predict motion Out There. Okay? All of that stuff was unknown to Mayan Indios. Yet they did put this quite sophisticated data into stone monuments. And now you can see where our new theories about Atlantis came from.

New section: Ancient math systems added

Essay: Ancient Terran Astronomy

In 1997 we started on this head trip concerning Spaced Out Topics. We figured out some interesting ideas based on data we got from reading some popular astronomy magazines. They were factual ones, not anything we commonly call the New Age or mystic ones. The data we have comes from very common sources but takes effort to find. We found out that certain famous Biblical numbers had a deeper meaning in terms of the "cosmic" because they were connected to many numbers found in modern science. This whole mission of mine by the way has nothing to do with astrology or numerology as they are known in the New Age. We think Astrology was the old name for what we now call Astronomy and that name change occurred circa 1500 because by that time we humans had finally gotten just enough data to make modern astronomy feasible.

Anyhow, here goes: we found that Sumerians had the numbers 144,000 then 666 and 216 to play with. They were common concepts and later appeared in our Hebrew & Latin holy books. Their **216** comes from the simple fact that, in cuneiform script, the digit Zero is often left out, but implied, and that leads to the perception that our modern system, called Arabic, has evolved from that, and it is now at least 7,000 years old. Here is my most vital formula: **216 times 666 equals 144,000.** That must suggest some strong, very strong, link between Judean and Buddhist kinds of numerology or math. They added myth and ritual to this. Which you are aware of by now.

Our basic link is between two Judean holy numbers, and one from Tibetan tradition. 216 comes from Tibet. Multiply this by 10 and we get 2,160. There are 2,160 years in each Sign of the Zodiac. **Our Moon is 2,160 miles across.** It all adds up. Yet how did ancient Orientals and Sumerians have such concepts in common? They traded with each other.

More data on Mayan lore. You can skip this. Are we being slack?

This Old World tradition of Science had other features: Decimal Math System but calendars/astronomy based on Dozen System... in other words, the Sumerians had ability to use complex systems of computation based on both "10" and "12" with these results: their famous 216 was based on the 2,160 miles of diameter of Luna. Each Sign of the Zodiac has 2,160 years in its Age exactly and there are 12, so that means we have 25920 years in a certain sidereal period based on the Vernal Equinox (Vega to Polaris) which we will call one Eon. Someone at that time began playing with math so that they must have had kilometers already. The Moon is 2,160 miles or 3,600 klicks in diameter. The figure **3,600** is called **Sar** in Sumerian and was also some long period of history. In fact they based dynasties on it. **That led me to deduce that a Baktun was really 360 years.** So a

Mayan Baktun of 360 years was convenient for me because it was another link to Sumeria: the Eon had 72 of them and that figure was also Mayan time period: The age of one human life which I chose to call one Tonal. It makes sense when we find that 7.2 Sars fit into one Eon. (Like 3600 x 7.2 = 25920.) But that was an error. We concluded that the Aztecs had altered the old Baktun of 360 years to 400 or that they had told some lie to the Conquistadors.

Maybe so, but not even needed. Here is an equation: 400 x 360 = 144,000 which means that 400 years by 360 days equals exactly 144,000 days. The whole thing certainly proves some link between Mayan, and Sumerian math, and we feel that the Mayan system comes from the latter because of this fact: **Mayan history began in 3114 BC.** Sumer existed fully for two millennia before that. Here is the other famous link: between the Hebrew tradition and the Tibetan: that 216 is known as famous "magical" number in the latter.

Actually the exact formula is: 216.21622 x 666 = 144,000. The whole thing comes down to a difference between the modern calendar of 365 days and many ancient ones of 360 days. The way to operate the latter is to have an official year of 360 days which begin on the Winter Solstice (in December) and that means Day 216 of our Eon, incorrectly as usual, stated by that nut and liar, Plato, falls on the modern July 23rd. The missing five days can be hidden in some bizarre festival time labelled The Timeless Zone or the Black Days Of Summer. Sounds evil, eh? Are you having fun?

This idea is alluded to in Mayan by having 36 months of ten days each making up one "year". This clearly operates without our Moon as measuring device. It seems to be yet another Decimal way of telling time and is convenient but arbitrary, since time periods based on the figure "ten" rarely appear in nature. Our star system has sixes and twelves going on more often (such as the orbit of Jupiter being circa 12 Terran years). Actually, to digress into Metaphysics, we can call Zero the nonexistent number and so is "ten" and any multiples of ten which means that in fact only nine real numbers ever existed.

This idea comes from Persian Cosmic Dualism, called the Zend Avesta. In it "zero" and "uno" contend as eternal enemies to begat all other numbers. That idea exists in Sumerian form but in very convoluted manner and cannot be explained here... you can get tons of data from the usual sources. So the Biblical number of 144000 days has 400 years times 360 days in it, or both Baktuns and Sars and is thus linked to Central America... but also Tibet via the number 216. The number 144000 is not of much use in astronomy unless it denotes 400 year period. My whole essay is about the links between Decimal and Dozen systems. This is an addon to Atlantis essay.

33 degrees West of Cheops

Some facts on Stonehenge:

The main fact here is the one that in many human Math methods down through the ages, two main systems have had their way. The Decimal and the Dozen ones. In other words, whether we use tens or sixes. The Dozen System can be expressed in any multiple of Numero Six and this is based on astronomy of our planet and its moon.

The decimal system is based as we know on our ten fingers which does not apply to the other one. So the two can be compared and used to measure our round planet... and also mixed with each other. That is where ancient Sumerians and later on the Greeks ran into controversy. One such expert was Erastos Thenes of 240

BC who had to reconcile our two systems of math in general. This posed problems when he had to use data from older cultures like the ones who designed Stonehenge after the Cheops

Pyramid was finally completed. This series of facts becomes vital as hell when we consider what Eratos must have known about the biggest monument of Hyperborea, that being the name of the UK in those days. It fits in with Number of the Beast because it is exactly 33.3 degrees West of Cheops/Giza.

The first equation is: $33.3 \times 20 = 666$
The second is: $33.3 \times 3 = 100$
The third is: $20 \times 5 = 100$
The fourth is: $20 \times 3 = 60$

You may be able to invent similar equations and I can guarantee that the numbers 666, 216, 33 and 144000 all refer to permutations of many complex formulas relating to our planet and how it relates to other parts of Sol System. It's only astronomy. So anyway, that stone circle in the UK simply gave ancient Man one nice slice of the globe exactly five percent of its circumference at the latitude of Cheops (30 North) or 5% of 666. At that latitude, one degree is exactly 60 miles and so that times 360 equals 21600 miles. Yes but at Equator it's 24901 miles which makes one degree there 69.17 miles exactly (or circa 70). Their main survey line lay at 30 North and was within reach of caravans across North Africa. We feel that the Phoenicians just found it easy to stop at Al Jebal in Maroc and draw a line north through Europe for their survey that determined the exact length of miles and other units but at that latitude. Here goes:

60 times 360 = 21600
33.3 x 10.810811 = 360
20 x 10.810811 = 216.21622 x 666 = 144000

There are some nice coincidences here and others that fit. The way all of these monuments were placed here along the 30 degree line on the Sahara desert allowed the ancients to play with numbers. The main idea was to compare decimal system to the other. We suspect that this line at Greenwich was chosen because it was close to the ancient marker at Stonehenge but offered an even 360 degrees. The main point is that the famous Tibetan number 216 is about 1.6, of 360 which is the ratio for kilometers to miles. Our planet is 24,901 miles in circumference at its Equator, yet only 21600 miles at 30 degree longitude. And the diff between those two is: 1019 miles.

Atlantis: We have three coords.
Giza is 30.0131 North. and 31.2089 East.
Stonehenge is 51.1789 N and 1.8262 West.
Total is 33.0351. Stone is about 33 ds West of Cheops.
Okay? That is the basic Masonic magic number.

Terran Cartography ends here with "Stonehenge"

More on Sumer:

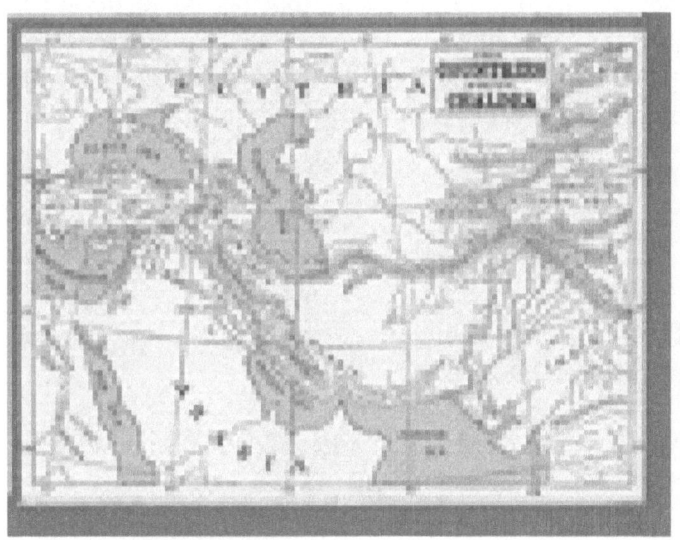

The Sumerian Alphabet

This one is seven millennia old and is actually demotic and not some sort of "rune" thing. In fact, anything like that was strictly primitive version or the origins of what was soon to develop into the modern Alphabet. In other words the Sumerians had by the Year 5000 BC at very earliest the world's first Phonetic letter system. We mean only that we have sounds, not words symbolized by our little characters. We mean not Chinese nor Nordic runes. So we are by now modern. They say that China had 3000 characters in their array and that Nordic runes had 800. Well that is too much. It is said that the Egyptians had what we now call pictographs. That was another set of letters of which each was one whole word. That is dumb and hard to carry out but it was okay as long as some decadent aristocracy was able to handle it. So here is the original Alphabet before the Levant or Judeans had it. It has 16 consonants with five additional vowels.

B...D...F...G...H...K...L...M...N...P...R...S...T...V...W...X\
par

Then we have the five original vowel sounds: A...E...I...O...U

And now we have the five letters that I have decided to remove as being redundant: J...C...Q...Y...Z

So we now have the reason that the Hebrew alphabet has mostly consonant letters and it seems to be okay to remove vowels. This shows again that Egyptian culture was based on Sumer.

Atlantis Essay ends here. Created 2005.

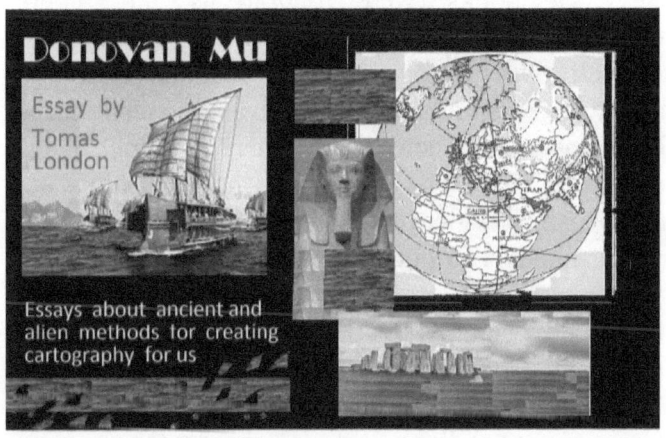

Fuck Atlantis Redux.

Intro: This essay was first written in 2013. Here it is minus some smutty language and more to the point, which is below. We had to edit out personal junk first. Really! We need a focus group.

The above explains James Redfield in his Celestine Prophecy. His books were fiction as the author claims, but based on solid facts from history. The story itself was based on the adventures of an idealist who went to Peru and was persecuted by local extremists. The hero thinks there is some ancient manuscript on paper that was lost in the jungles of Peru when Spanish invaders killed natives. That was long ago. He thinks this document proves some kind of link between the Incan and Mayan cultures. It is a story of the same old arch types. Here is the bottom line: The Mayan culture was in some way influenced by Sumerian culture directly but it was back in 3000 BC at least, if not earlier. Some Sumerian words seem similar to Mayan ones. Later on, the Egyptians must have sailed up the Amazon in large fleets with large ships and so, eventually contacted and traded peacefully with Nazca natives. Then over millennia, those two cultures mixed until they became the Incas. So since Egypt evolved from Sumeria, both empires (from the Old World) somehow influenced two regions of the New World. But you see, Sumeria first went to Central America to settle and stay. Much later on, Egypt was to appear in the Andes to settle there.

Bottom line and more to the point: This is where I got my ideas about Atlantis from. They were basically by intuition. Actually, over the years some strange alien handed me some clues. Which was cool. It was a cosmic joke. But see, it was still a valid idea. She reminded me of ancient legends of Atlantis which we decided were propaganda made up by Plato, to cut it short. Seems he created stories of two entire continents sinking under the ocean. (That would make one giant World Ocean of the Pacific and Atlantic. See above.)

Well it did not happen that way. We mean according to what Plato said. In reality, only a cartographic grid pattern recorded on a few charts were lost. (That happened in 410 Ad with the Fall of Rome.) Most of them were secret, for military reasons. So our two island continents still stood above water. I mean physically very small areas along the coast were covered with salt water as land slowly was covered with water as polar ice melted. That was of course only between ice ages. That explains any seawalls found at Bimini. In fact, it has been called part of some Olmec harbor. No mystery here.

Yeah, okay, so yes, over millions of years in

geologic times those Ice Ages came and went. So the Sea Level rises and falls. That was recorded eventually by some ancient cultures as Atlantis sinking and rising. But not the whole thing. Maybe just... a few hundred feet. That will cover up some stone temples built long ago on some beach on the shores of Mexico, etc. You can see the wheels turning. So the water rises and falls. So what. And we can even tell you the exact period: 25,920 years.

Now to make it short and sweet: For some reason it is claimed that these Natives did not invent the Wheel nor use it. Nor did they have nor use Iron nor Steel. As we know, our steel industry existed since maybe 1,300 BC. Before that we had copper, bronze & brass. Note that the making of steel is a sharp step above any previous levels of metallurgy. Blast furnaces are more complicated than what any copper industry requires. Nothing to us, but a major step forward for our ancestors. Both levels of technology were invented in the Middle East. What took so long? In the Old World they were stuck in the Bronze Age for over six millennia.

Again, so what? We can suspect that the earliest Mayans and Incas did have iron but only the rich. The poor had to use copper or even just stone, wood and bone for tools. Those people were as smart as anyone. They could have invented their own steel industry, but never did. Or if they did... they lost it. How? Over the years on the other continents, which were not as isolated, iron & steel was used and by all races and ethnic groups. It was universal. That is what our modern science tells us. They even had wars with each other and many lived in great empires. But they never totally "bombed" each other into Stone Age poverty. Why In Atlantis?

The author thinks there were aliens who landed on a regular basis to observe any hi tech stuff being invented, such as any steel industry. If it did it was destroyed. As some weird idea, we can take the Spanish Invasion of Mexico in 1492 as an example. If any of those Atlantean (their real name) nations had any steel then Spanish armies destroyed it. It would be confiscated and melted down to make into armor, weapons and other stuff they needed. That Spanish War Machine had a use for steel as well as for gold.

This concept is simple. For some strange reason aliens allowed Old World culture to use various devices like steel but not in the New World. It was probably some experiment in genetics or cultural relativism. We do not know what aliens think like so why bother to speculate? The clues we have are all based on astronomy and other modern data which if combined with what our own history tells us, yields spooky results. It is time now to respond to some hints. Like if on ancient maps anything can be any size, then Iceland can seem as big as Greenland (which was once called Ultima Thule) and even Baffin Island, Labrador or Nova Scotia, being various kinds of peninsulas or islands often covered by icecaps, can be about any size. All of them can be called Island Continents. So that term was just some stupid label valid only for ancient people. They had simple minds. It was not until well after Columbus that it was clear that Atlantis, aka, uhm, North America was much larger than any island near it. Same thing for South America, as Mu. So there. This is a view common to aliens and us modern humans. That is what should be. They all mention a certain big cosmic event from 600 BC as major point. The Mormons did. James Redfield, whom we have mentioned above, hinted at this event, as well as other famous New Age writers who dabble in the Ancient Astronaut Issue. This may sound incoherent, like some rant from the 1990s, but many of these writers come across as if they waste most of their time swindling. They are usually not aware of this. Yet by now, both lying and debunking are out of fashion.

What is the bottom line on all of this data, which came out from 1965 to 1997? (I am starting with Clarke's book and ending with anything the 1990s New Age produced.) Here it is: About 600 B.C. the last Assyrian king, who ruled Babylon, was defeated by a coalition of Judeans and Persians. Since this regime was the last gasp of basic Sumerian culture, we can compare the Fall of Babylon to what happened in 1991 when our Western Coalition started another regime change on a similar villain. That was the First Gulf War. Eventually, Saddam regime was deposed. See the Enuma Elish for similar events.

The results were that an ancient regime collapsed. Among other things, records in cuneiform were made available. Those must have contained huge amounts of valuable data. Finally the world had full access to really ancient astronomy. Around that time, an expedition was organized by the combined powers of Judea, Egypt, Greece, the Levant, Persia and other nations who had an interest in good science.

Once again, they wanted to prove that the world was round. Also, they wanted to find out the exact circumference of this globe. To confirm all of this detail, they combined funds and sent a large fleet of ships to Atlantis. They landed in the Yucatan, traveled over land and contacted the natives in a peaceful way. (Maybe this sound a bit wimpy, so it was left out of all records except for what the Mormon Faith has to offer. Right?) Eventually they reached Teotihuacan, which was a basic replica of what we find at Giza, and confirmed their data there. We assume they sent only a small force back home to deliver the evidence. Maybe a few pieces of gold and some charts. The rest of that crew stayed behind forever. They merged with local culture. Needless to say, many Old World customs like a steel industry and Monotheism vanished. They went mostly Native. It was a question of adapt or die.

Now flash forward to 1492 and then 1519: Assume that this Spanish Conquest was recorded by modern history mainly because it was just another good War Story. It was dramatic. At least these people finally achieved something. They took horses, gunpowder and steel to the New World. They also took things like tobacco, corn and potatoes back to use. What else? Among other things, they claimed that the Natives, meaning the Mayan and Aztec Indios, had no way of transporting people and goods over long distances. Like, they had knowledge of the wheel but did not make use of it for common farming, industry and trade. They could easily have built many carts of wood with wooden wheels. They already had toys of clay for their children.

These were made of clay. The only reason for this is, that, of course, even more of these toys, made only for children, were of wood. But in the tropical climate those objects just rotted. Decay. Likewise, any carts of wood, with wooden wheels which the Indios of Central America were capable of, either rotted away or were burned as firewood.

After 1519, the Conquistadores must have confiscated any carts to use for their own military. Why not? And forever preach about those Pagans. One more concept: While carts of wood, with wheels, were okay with the working classes, and even with the more wealthier merchant class, assuming the Mayan Kingdoms had any class system, which they did have, why allow any wheeled traffic into the small confines of their Holy City? Why not ban them under Mayan Law to suburbs where poor people lived in wooden huts? That is where their industry existed.

The above rant is what any modern Sociology writer might have said, under different circumstances. The book called "Collapse" by Jared Diamond comes to mind. This concept has nothing to do with Alien or Cosmic influence. We did wish to explain such things, but for now, let Earthian concepts rule. They are basic. Maybe later on with cosmic. Do not blame NASA and similar outfits. They get paid to do square science in the square way. To apply to Congress for funds. To lobby. You know that. We know that. Now use imagination. Try this: In 1993 one American New Age author wrote his book called The Celestine Prophecy. It was meant in some emotional, very personal, inspirational way. Which was groovy. But what if we sent some normal, materialist archeology team to such places to find clues? They might find that, in order to make surveys of Planet Earth as to circumference and other astronomy, some team of Egyptians and others in 3114 B.C. actually went past the Gates Of Hercules, sailed West near Senegal, then hit the coast of Bahia, Brasil, and established a small base? See the 2017 movie about Fawcett, "Z" The Lost City. Then they went on to the Yucatan and ventured inland. They set up another base camp at Teotihuacan. Then they confirmed whatever their database. They left some of their party behind to build markers of solid stone or brick. They returned to their kings to report their pure academic findings. Then the Natives of the New World took over. Astronomy was welcome. Now what about aliens?

We say that event "caused" Atlantis to sink in ancient myth. (So in that case, when did Mu sink?) That handed me a big clue. So if one can find out what they are trying to say, and then even codify isolated comments, then you can make sense of the whole concept. You need to have the mindset of the modern military or an academic to get the point.

Then you need to try to analyze aliens who may be millions of years ahead of us. The Moon, aka Luna, is in our hands now. In 2018. We can land there again with armies and bombs. Who cares? Them? Get the drift?

The end.

The above essays were not fiction. Based on data from public sources. Published by Tomas London and Amazon Inc. on February 27, 2017.

Cydonia My Love

The same old fuzzy photos. From Daguerre in 1830 in some small French town to Foto Shop in 1985 in San Jose and then to the immensely detailed digitalized wonders of AD 2084 on a NASA Starship travelling to the rather obscure star Epsilon Indi. We shall not get into the full details of it all. I mean; like how can I possibly explain fully the way we have finally achieved Nth degree of sophistication in this era? Why Anno Domini? That was an old way of dating. Who uses Common Era anyway? It is actually now the Year 21,394 of the Eon Platonic, according to real astronomy. Cindi feels strongly that any aliens if they exist must use the same system. That is, if they have any connection to Terra.

Cindi likes using the old Sci Fi terms; in a mere playful manner to keep her mind active. Sure she hates to waste time like anyone else but they often had no choice on some ships but to fill my head with trivial stuff that serves to keep negative vibes out. Some of the other crew members really think "demons" exist and can possess them but that's their problem. Cindi Nord is her real name. She is now sitting at a steel table in one cafeteria of a huge fusion drive ship. Far below, plasma at 30 million Celsius is being blasted out of giant nozzles and their speed is already into circa 60% of "c" but she does not notice because this ship is being kept at a mean acceleration of one Terran "gee" force. She is a petite person of 28 with lots of curves and a cute little face. Like most people from the European Union, she has short hair which in this case is blonde. She wears a yellow tank top plus blue bottom of cotton according to NASA rules and the mores of her Captain.

Cindi is shifting through all the data on Cydonia she has. These photos of a controversial part of Mars may be chosen to appear in Solar Times magazine. She has to choose the most dramatic views. Some are still on older types of film with chemicals; others are pure hitech. All are high resolution. Actually none of these photos can prove anything. They are just decoration. Men long ago entered these ruins and had found the remains of alien culture. But she had work to do. Her hands (which had no special powers but were nice to look at) sifted through older 11 by 8.5 glossies of Pyramids of Cydonia. They were from various missions. NASA from 1976 to 2008 and ESA from then on to 2040. And also the Face. She still had a famous book from that area, all 500 pages stored on her new ABM as in American Binary Machines Superpad.

She spent long time spans looking at the old glossy photos with her magnifying glass, but also used her notepad to "zoom" her "pix"... and so on. This was just a pleasant activity. So tranquil. Her mind wandered - until the minute her tall friend Alice walked by. Just passing by. This was actually a Starship called the Johnny Appleseed. It had nothing to do with Mars.

Alice walked in to have a look at what was going on

in this place. Cindi seemed to be in a trance. Over the weeks she had turned this large empty dining hall into her own private study and/or chart room. As this ship was now moving along at a steady 25,000 miles per hour, which created a force of one Earth gravity. This force was directed straight down so they could stand upright. It was a way of creating artificial gravity without any rotation. Thus by 2084 they had gone ahead of what we can see in methods of "beating" conditions found out in Space, even far beyond "2001" Space Odyssey. Like that rotating cabin on the Discovery. Now they had both some gravity suitable for Terrans and forward motion. Alice was here by chance. They were still trapped within Indi System but also relativistic travel. Seeing Cindi here wasting time annoyed her. Yet that was on a basic human level. Nothing to do with galactic politics.

By the way, none of all that was known to Biederman, nor the rest of them. It was honestly beyond her own control. She was free to do some research on board this giant but almost empty ship. Designed for hundreds but with a crew of only 21, it counted as a ghost ship. May have been controlled by something unseen. Alice had to find out. Maybe this was only called her karma, as in her own personal mission in Hell.

"Why are you here?" asked Cindi.

"To piss you off." came from Alice as she stood on rubber floor just behind Ms. Nord. They tended to become abrasive towards each other on occasion. This was almost sexual. They tended to behave like a typical gay couple. Which was often bitchy. Sex was not really an issue in the 2080s anyway. Often anger was.

Cindi wasted time explaining herself politely. But she still did so valiantly. Alice had a cynical idea. She said, "I feel you are wasting time. Not to say that we may have better things to do, you know."

"We certainly are not here for sex. I still have a silly thing going with Fred but he needs to get his mind off what we are doing here."

"Does he - in any serious way - in anything - get anything out of this? I feel lousy about it."

"De nada. Fred is okay with me. But has no idea of what we are doing here instead of that other star, not that it really matters by now. Right?"

"We have no fucking business with Centauri System. The fact that Jerry had Centaurs live there was a dead giveaway. A giveaway. I mean, think. In that old hard kind of roman. There. I've used both German and French words for un nouvelle romance." Both started to laugh. They began to relax.

Alice said "Here. Will be nice and explain what this whole shebang is about. I can even get poetic about it. I can even quote Ozzy Osbourne. What did he call Mars in his gloopy song? On his debut album back in 1970?" Cindi had no idea. She was familiar with hip hop and other stuff. Which did not matter. Music to a child of the 21st Century was no big deal. Even Alice understood that. So she tried to link ancient hippie music to even more ancient mythology. Which as of 1492 must be obvious.

"Look: Under those pink, yellow, salmon, creamy, beige, tan and frozen sands of Great God Mars, here!" Like in that song by Ozzie Osborne. Her finger pointed at angular shapes. "These things were built in fact in ten million BC, and that is so long ago that our own calendars are so alike it becomes academic. They are a full kilo meter taller than they seem. Yet, my abductors tell me they are the most recent structures on the planet. Look at the edges of them and I'll return to love thee some more thou small Being." Alice got up and marched out. She was in the same outfit as Cindi, and barefoot.

Alone, Cindi's mind was relaxed enough to conform to her pilot's attitude. Chain of command? After all, she was only a mere Flight Engineer. Cindi felt deep resentment. What do we have our comlinks for? Our 21st Century mastery over matter! Eh? What gives! So she dialed Biedermeier in his Command Post to kill her pet peeve. "Fred. Yes, I mean Captain" she said into her horn, "Need to raise issue of Alice and her recent overbearing..."

"Fred ain't here." came a female voice in answer. "Fuck Fred" said Cindi. A second later, Annoying Presence made her return. It sat down again and - gave an encore. Cindi screamed across the table, "What the fuck do you want?"

"I want to save you time. Aliens have been living on Mars for a long time. We still do not know if any life originated there. Yet for billions of years now, beings from elsewhere have been landing there. By 2010 our own rovers began to see signs of this. Those pyramids contain bones of various Races. Some are Archaea from places like 61 Cygni and the Project Oz stars. They like the cold biome down there. I get these vibes from them - " Playing along, Cindi ranted, "Fuck you! So what's with Mars? Speak your piece or fuck off!"

"So I love you all. My poor Terrans. Here, I'll tell you tales of your Red Planet like your Good Doctor never did. Like -"

"I have had enough!" Cindi raved, "You hate them. You called them that six letter word! On TV. In private. Like in the washroom. Or even my shrink's office, and he has to edit your nasty mouth from his tapes. That was his job. So the American public cannot hear your xenophobic junk. You are the greatest waste of oxygen yet. You seem to hate slaves of some kind. Right? "

"That is very true. Earth is the real Planet of the Apes. That has nothing to do with beings from Out There. OK?"

[Pause.]

Author's note: To barge right in. What Cindi does not understand is that Alice has a Platonic kind of "love" for her fellow humans in her own way. She is obviously turning into what Dave Bowman was turning into. Only this story will try to explore that concept in more detail. Okay? That movie came out in 1968. In those days We were still wondering about sexual mores. That stopped less than a century after their "revolution" of the time. Hence we still have echoes of hetero, homo and other erotic vibes floating around in this genre. It is usually Platonic. It started in Victorian times. We had some "ribald" lesbian humor in this story but that fell flat. This is better. It sticks to the Original Odyssey. Okay. A point: Some critic once speculated about an affair between Frank and Dave with HAL 9000 being being effeminate. Ha, ha. But that kind of thing cannot ruin the story itself. Maybe this is closer to the point.

[Pause]

"Sorry for my crass humor. That sort of thing was okay back in New Worlds magazine. It comes from the New Wave fad. I was wasting time making fun of those aliens I had to put up with." Cindi relaxed and said that at last, the talk was improving. It was as if sexual innuendos had to be made first. To break the ice? Then more abstract ideas could be risked. Now we have a better life to live.
"Do me a favor." said Cindi.
"What?" came from Alice as she sat down somewhere in the room and waited. What a mood change. It was all ears now. "Let me tell you a story from France. It starts back in 1830 with photography. Then ends with radium discovered in 1905 in the same nation." Cindi was calm. Now that some kind of mind was possible in their talk, fleeting traces of aggression sank to zero. Alice sat there like a robot placed on hold or hibernation. This confirmed much.
"Tell me your story."
"This is good" said Cindi "As of 1830 it became easier to record visual reality with machinery. Like the camera. It was almost as if the Art of Painting had become redundant."

"Oh, god! You thrill me. Do go on."

"Here it is then. Painters who used to make some pretty penny with normal art began to lose work to photography. The end result was poverty. They lost status. Then along came the Great War. By 1920 the average person, secretly or otherwise figured the word 'artist' was synonymous with "bum". An unemployed person. Right?"

"Yes. I agree. But that was not sexual. It was all just a matter of the economy. I can get into that." Now that they were on a polite level, Cindi felt better. She said "Listen. As the 1800s went on, artists had to innovate. No longer was it enough to paint normal reality, which had been lucrative before, they had to turn to direct ways of expressing their feelings. Their inner life had to come out. First we had to endure Impressionism, then Expressionism, and so forth. Art became progressively more abstract until by 1920, it was all abstract. Picasso was talented, but society was so fucked up that he had to do 'garbage' to make money. He was probably not happy with this. He was also considered to be mentally ill. Comments?"

"I agree for the most part. Picasso, like most creative people in those days, was seen as a radical. A freak. But of course, in Western society, all radicals are considered to be mentally ill. Hence all artists. It's implicit. I mean even now in 2085. Just for a joke, how about Toulouse Lautrec?" Said Alice.

"Don't know much about him. But I did study another case. This man is more interesting. Vincent Van Gogh. It was said that he cut his own ear off in a mental institute then painted himself displaying it in his hand,

as if he were proud of his insanity. That was not so. After checking him out on Internet, I found out the truth. He was just drunk at the time he lost his ear. He and another jobless jerk lived in a cottage together and tended to abuse the vino. One day
they had a fight. His so called friend cut Vincent's ear off. Also, the cops were called. Later on the media claimed that both of them were insane. Maybe they needed to get a pension." Cindi paused. She remained at her table. The she said in a calm tone, "Media. What am I saying? Papers only. No radio nor TV. Gossip in a small town. Personally I think Vincent's friend or roommate was arrested and in jail facing a serious charge of assault with injury. Mutilation. He made up this story of how Vincent was in some local asylum, where he had psychotic fugues and one day cut his own left ear off. But I really think he told lies to get off." She fell into reverie imagining that clever thug, later, as excon, telling jokes the mentally ill to drunks in Paris taverns. Eventually his Bohemian and Demimonde *amies* believed him. The point was to sell Art. Create a sensation. Bonhommie.

"I've never heard of this before. It fits in with my own notions of mental illness. In the 1970s, most of us did not pay attention to that issue. We tended to either conform or rebel against society. We used to believe in Good and Evil. That was it. You know, I am glad that I even provoked some reaction. You have really helped me focus on my vision." That was sincere in a quiet way, coming from Alice. For once. Both girls were just sitting down and chatting. Cindi said "I can see that I was wasting time on those wimpy old Mars photos. But I do have a job still, it's even on my roster. I have to process any valid evidence of alien life as we encounter it. This stuff is backlog from ESA missions even back to 2020. We still did not assess tons of Mars data, and you should see the amount we have on Jupiter! But what am I to do."

"A clue, my friend. Was up in Control Room a few times too many today. It was empty with door open. Fred is hiding somewhere. But it does not matter. I can see his logic. Like, now that all Drive controls are frozen, and we are in - well, you know what? This tub is going nowhere. I can wander in and out of that place all I want. Just knowing he does not trust me, but what can he do?"

"What or who is controlling this ship? Be frank."

"Some alien. we have to wait it out." said Alice.

"You see, I am still on the case. I was trying to get a fix on this critter by looking at Mars photos. Alice, do you realize how many races could be involved? How many empires?" Alice shifted in her seat. She played with tools. Then she said "I am not a god. It is clear to me that we humans could not even prove aliens exist until 1947. Since then we gave up looking for them. The best of us tried instead to figure them out. But that cannot happen until we actually leave the Solar System. Now that we have tried to explore another star system, and failed, we are part of their society. We on this ship are no longer a part of Earth society. We are no longer Earthian but are being absorbed by them." Cindi felt a cold wave of fear flowing up her legs. She said, "You and I know about it. We can take it. What about the others."

"Fuck them. They have to take care of themselves. This whole thing is some kind of selective process. Evolution the basic way. But one factor I can deduce: Radiation."

"What do you mean."

"In the first place, we have had it easy. Last year we ended up in a simple system with just one small star and a few basic planets. *Nerus* had an ecology, okay for us to live in. Even that was very good luck. Also, it was devoid of any life at all. Not even microbes to kill us with illness. Nor sentient types to use weapons, etc. We even had interesting ruins to explore, which was the case on Mars as well. Okay, Cindi, I can see why you had to waste time on this old shit. I cannot blame you. I can see your logic. I can even dig the point of Fred not being on duty where he should be. Now what I need you for is some more deduction."

"Okay. A long shot: Had we hit Centauri System, what should have been the result?"

"I can play ball. Man, we're both super with NASA! Like real Rohan freaks. It all checks out! That system has three weird stars in very freaky orbits. Why not assume that their white dwarf is that Demon Star, with an actual medieval world as part of that whole fiction setup on it, and Tran is for real? Eh?"

"Sure. Those old writers must have been on to something within Known Space. It must add up on some level. But in any case a system like that is hostile to our kind. We'd be better off just in terms of biology to try for this system. It works for us. Or maybe on of the Ozma pair. Fuck Centauri I would say!" Alice agreed.

Both were happy with each other.

"One small comfort," Cindi said. "This force may simply be some cosmic mechanism that made us do all this. It can, okay, even *invade* our own computers. So Fred is smart to take it easy. Why waste time yelling 'Khan' in anger? What is it? A test? What would such atavism prove? Fred is a meathead, sure. If they had bodies, we could just waste 'em with tommy guns. Or my mace. This has to be some vastly sentient force field guiding us through centuries and voids for some basic grounds! There was some radiation down there. If we go few hundred years into the future of *Nerus*, it will be close to a perfect world to live in. Some of us could start a colony if we had to. Even if we are stranded in the future."

Alice walked over to the food counter. She calmly poured fresh coffee from an urn. Then she looked for snacks. Cindi remained at her table in a trance. Soon Alice returned to that Cindi and sat down to have a meal. She said "After all my time on *Xiotan,* I can guess who is doing this. There's war going on. Warfare is her basic motive. These beings to be frank, are all hardwired into it. That has nothing to do with us. On our own, we would have been creamed in all this conflict between many different empires. If something has to guide us to the right place and time, so be it. I am happy to have Fred hide from Zeno and to see you sitting here with your nerdy work. It keeps us sane. They may still respect our minds. They want us here."

"You mean I figured out stuff about them even you could not get at?" Alice had to agree. The day wore on.

END OF CYDONIA MY LOVE

By Tomas Londan, January 2000.

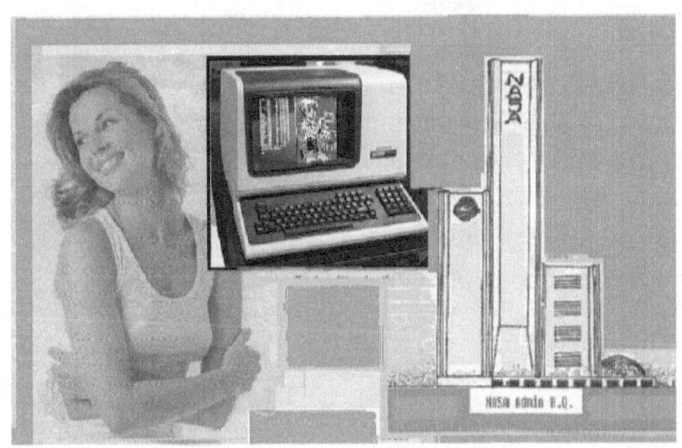

Rosetta Stone

Alice was on the horns of her old dilemma as she tried once more to appease her unknown boss. All she knew was that for two long years now she had only one enigmatic assignment. It was to please her boss with her personal best on a worn old Apple II sitting in some lonely basement room in an obscure office tower in downtown Houston, Texas.

This deal had sounded weird from the beginning and had not once improved since October of Anno Domini 2000, which was by now almost 24 months ago. Many strange things had taken place since then, yet this stuffy room had not changed in all that time. Alice had to spend eight hours in here from nine to five every weekday no matter how hot the air nor how windowless and full of carbon dioxide this veritable tomb tended to get... to her senses. It really was only seven by ten feet and had no opening besides one steel door. In fact, one of the hipper brokers upstairs had speculated once that her office had been used first as a vault for cash but that was back in 1930 when this block had been built. Yet we digress.

Yes, her job. She had nothing to do but sit and monitor an Edsel of a computer (we mean it: this thing still used BASIC and DOS). That was about all that could be called obsolete. However her actual purpose, was as advanced as could be, even in our 21st Century. It was to monitor any message coming in on her primitive PC. It had no sound nor CD-ROM; just big green letters on a small black screen. This klunky monster of white plastic sat there on a standard desk with one metal chair in front of it. Then she had many boxes of floppy disks crammed with the results of our State funded experiment in bongo drumming. Yes, floppy diskettes. Back in the Stone Age. This used to be called Project Bongo. As if the people of Texas were slacker natives on some island in the Pacific, trying to impress Captain Cook. Then, after September of 2001 the name suddenly was changed to Project Rosetta Stone by her military bosses. That concept led Alice, today; down a familiar path. Her mood was confirmed at 9:20 when another nice young dude came in to ask dumb questions. Communing with unseen entities was not on the minds of men who strayed sown here. Not usually. Alice looked up and said, "Hi! In answer to your... I am only 25 and single and no I know thee not. Thou art welcome."

"Cool. Well, I am actually not in that mood today. I already have enough... cookies. Never mind. Uhm, how much does that earn you? This act of yours is simply tres boring for mio. All of the guys up there said the same. May I ask?"

"Sure. Only I can do this. I'm actually trained to do exactly what my USAF manual says. You couldn't function at this so I make say, 34 grand per annum."

"Not bad. Wages for simple programmers are less than half that. Of course I make a good bundle. No offense, but I don't envy you. Now I can see how you can take the boredom. But what's the point?"

"What do you mean?"

The broker told Alice that she kept her door open most of the day and only locked it after hours, as if all of those records of an unseen person typing out some useless message once in a while when it suited them mattered to any human here. It was not even faintly romantic, like as if it were an old teletype in Dr. No or another one of those Sean Connery movies, or say The Man From Uncle, banging out reams of coded orders from gangsters in Hong Kong. It was sentimental in the least, just dull and depressing. Alice had made friends here in downtown Houston and had not even left the army. She was paid far more than any Private and even had the privilege of wearing civilian fashions. Today she had on only a mini dress and bleached short hair.

"Well, the point here is merely to answer whatever the Apple can type out at random. Then I have to store all of it. Both whatever I talk to says, and - my own talk too. Real simple." She drawled that out and stared up vacantly as if she were adverse to speaking to any other human being. Both of them started sweating in the heat, which had to be a human enough reaction for anyone. The young man had nothing against her now. Except for this very familiar nagging voice from the back of his mind: Time Is Money; it is money - that you are wasting. And tried one more path of inquiry by asking, "Who is your boss really?" Alice smiled and said, "I work for the US military and we just rent out this space here. My cubbyhole. So I do spy stuff but my real boss is not known. None of us have seen his or her face." Floored the broker. He backed up and wondered about her sanity. He avoided the obvious in conjecture and merely grunted, "Like Russian or Cuban or Peking? I now have it figured out. And I gotta run soon. Break is over. One more question: what happens on washroom or food breaks in here?"

"Too easy. I lock the door and leave. This PC is programmed in its pea brain - fast and stupid - to record The Other Side and that is okay until I get back. See?" She in turn had nothing against other humans at all. She welcomed the odd talk with Real People.

"The Other, eh? Weirdest Ville. Oh well." said the Broker and returned to his work none the wiser. Alice went back to staring at her screen as if nothing had happened. The Other had not sent any messages since she had opened up the room half an hour ago. After minutes she finally saw the familiar green symbols appear. Type on screen: C:\\> Hello. Have new job for you. Ready?

Alice: Hello yourself. Just had nice chat with nice man. Why so abrupt? Why hello, not good morning?

C:\\> Am being Orwellian today. Me my fave joke for you and your, as we all say... nice men. Ha, ha. You are so funny. Too bad this clunker does not have mikes. But I like type interface. Keeps things on my own mysterious level. Remember?

She: I do not follow.

C:\\> Lost you there. I like to have the same ambience as back in 1952. The losers came in 1947. We came later. Got that?

She: "I sort of recall that. So what?"

C:\\> The basic idea as of our arrival was to allow

your leaders to put the pieces together. It is one big cosmic game. We will have some things to do here and not all on your continent. Policy as it is invented by us and directed at other ones. But for today, Alice never mind that bull. Here is your question for today: How do you feel about the way the world is going in moral terms? That bothered Alice. She was stopped cold by the moronic tone of the last few lines from the Other - for that was her name for that Thing or Things she had to deal with. She had, among other things, never in two years been able to figure out whether They were just one or more beings. But more vital in terms of logic, why suddenly this obsession with morality? They had never stated this subject so openly before. All typing had been about matters of technology, money, even how to wage wars on this planet. History, yes, but never this tone of inquiry into values. Alice hesitated.

C:\\> Since somebody musta cut your fingers off I will give you a break. Take it easy and answer me at 9:40. Over. That was it. Her screen went blank and stayed that way. Alice took the time to rethink her position once more, for the hundredth time at least. In the first place, it was 2002 and she personally had the good fortune to work for the best of all Terran air forces. Their worst running joke since 1964 was this: What do you say to an organic tiger - not the paper kind? Simply this: "Hi there!" Well you had to have seen the famous movie by Kubrick. Alice just knew

in her bones that her human bosses could easily say

"Hi there" to any Visitor from the Void. That does it!

She figured. Of course she had been spending the last two years talking to aliens. What else could it be? $68,000 spent on her so far for this? To create some cheesy hoax?

Well, it may be debunkable even so. Alice privately figured that the USAF may conceivably be making up all this "saucer stuff" just to justify a huge Defense budget. So what? To be honest, Alice felt that whatever she communed with was not human. Period. Now to the relevant part of her problem. On the face of it, she had to do boring routines each day. It must have seemed like a meaningless and even depressing existence to any casual observer but one point they never got: It was actually fascinating and vital to the defense of her nation. It often happened that some top dog with whole rows of stars on his epaulettes came by for a while during the day to offer his **tonic to my troops** quietly. It was never commented upon.

Alice spent whole days with nothing to do but read books and waste tobacco just to read and "Store" some enigmatic message that in some manner made sense to her Brass in Denver and inside the Beltway but to the casual observer seemed more like Numerology. At 9:40 the screen lit up to allow green letters spill out. Her message for now went: "Hello, Alice. Have you been giving my question the proper consideration?" How blunt. These were indeed interesting times, as her waiters down the street used to say. Dealing with the Others reminded her of many corny and outdated ethnic jokes. They were different, after all.

They must have provoked jingoism before. Alice used shrewdness to go for subtle speech. Revealing subtle and kind thinking. She typed "I feel pleased with moral values within my society in general. I don't know what is happening elsewhere but where I live we try to treat each other as kindly as possible even in these difficult times." Let us see what they make of that.

She expected some sort of cliched response. Knowing this urban area, which was large and lively, she already had several silly ideas lined up. Most of them had to do with crime of course, but that was only due to her long experience with the mentality of the Other. She was used to concepts from these aliens which to any normal person had to come across as being simply moronic; as if alien visitors had no way to gather data except by the mass media. The Other had the boring habit of speaking in headlines.

Alice continued by claiming that she tried to get along with all of the people in her life, which again was vacuous. She then gave several examples. This went on for the rest of the day but to her was welcome because it kept her busy. In fact, she was so busy thinking of ideas along such lines of moral reasoning that her whole working day went by faster. Since she seemed to be busy, the public ignored her despite her open door. Finally at 12:27, just before lunch hour she paused for the Other, who had been limiting responses to short words of praise and encouragement.

The Other said "Why do you keep the door open all the time when you should be concentrating on proper answers to my messages?" which got silence from Alice. She felt insulted. She had never uttered the truth before, namely that she had felt so threatened by the presence of aliens that she would have gone crazy. She had felt so within her first ten minutes there. As far as she was concerned, no human could tolerate them. No wonder they needed hitech interfaces.

From 12:30 to 13:30 was lunch. Alice always left for that, but course locked her door, which was marked simply with a Number Five of shiny brass. When she finally came back she read the expected answer and it was this: "You lie. Please rephrase."

Alice decided to go along with this jerk so she typed: Okay you win. Have it your way. I now feel that life on this planet is worth living - not. In fact why don't we start total war.

C:\\> With nukes? [It was the Other.]

A:\\> With nukes if needed, yes.

C:\\> Why? [That was predictable.]

Alice felt like typing "why not?" but remained in control since, by experience clever words got hateful answers or silence. Her boss in Denver merely told her long ago to keep the dialog going as long as she could so that they had large database to work with. Their purpose was to analyze any alien mind they could commune with, no matter how tenuously. It was for vital strategic analysis. War loomed as of 1947 and grew larger as threat to Humanity. So Alice tried another tack as sailors used to say, as in "When did we first meet, alien?"

C:\\> Not that I care, but on August 24, 2000. Do not use that word again. Call me the Other always. You know that we of my kind in those days told your Brass to set up this room and use the oldest piece of junk for an interface. You recall we used radio then.

A:\\> You have made me happy. And I shall live long and prosper with mucho dollars. Where does this line lead to? Some secret cave in the mountains with some wee critter at the other end?

C:\\> It leads to the Internet. Now tell me why you lied - that is all we need to know.

A:\\> Let me be frank. I decided to suck up to you. Would you rather have an honest answer? Make up your fucking mind.

C:\\> What did Stalin say to the Pope?

A:\\> Dumb. Look amigo, we may not all be human but we are still fellow beings. I am relating to you personally and still wish with all my heart to be logically consistent...in either pleasing you or dissing you. Right?

C:\\> Why do you lie?

A:\\> I believe in the Brotherhood of Man. This planet, if all of us work together, will end up with five billion happy human beings and may even become an Ecotopia.

C:\\> I am sorry but that concept is not valid. No matter how positive your outlook on life may be, we aliens do not care about you and your race at all. If we ever choose to invade Terra we will - we already control your society. This tool of yours is not just another Rosetta Stone, but an Interosetor; the same device of alien origin that appeared in the movie This Island Earth. It ran back in the 1950 and sort of matches reality.

A:\\> How so?

C:\\> We are various races and so far ahead we regard you all as animals. The entire planet is one concentration camp. It will be best for you to fight wars with each other since you cannot fight us off. That was the last message from The Other for Alice. She thanked her Overlord [?] and then took an unscheduled break on her own initiative. It was even a smoke break which was illegal and therefore unusual, at least in this age, and not decades ago. Alice thought casually to herself: I will get back to my friendship with my overlord again but later today - also we tend to have a good time communing with each other. [Editor: So is Arthur C. Clarke a concept?]

The ultimate fact for her was that within mere months one alien entity had enticed her into accepting Them as her intimate friends and not her five billion fellow humans. That did not bother her. It was a new fact but one that has never disturbed her race before and would not do so in the future. That was the basic message from all of our Interosetors. **The End.**

Rosetta Stone, 2002.

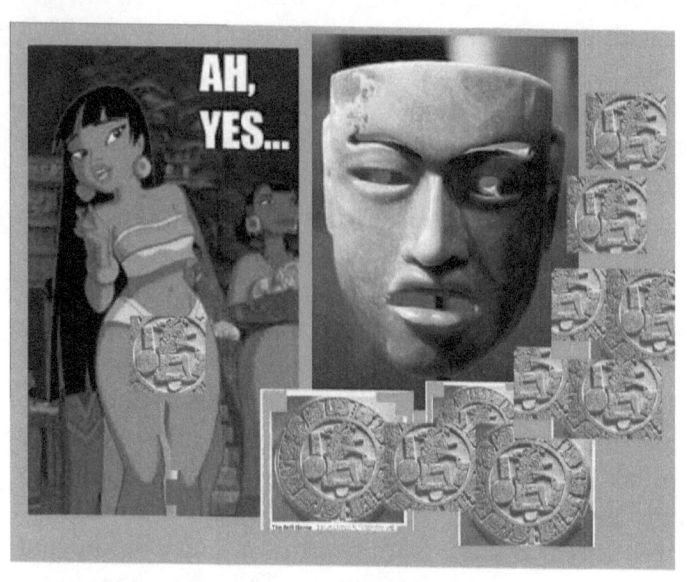

Rain

by Tomas London

It was something subtle that happened only a few days ago on a very quiet and wet Monday afternoon in some small town in North America. In fact it reminded me of an old film called Silent Spring. Actually, since this was for real, as in one 10 minute slice of my life, it was in color but the film was not. Suited my theme. But anyway it was an overcast day in November, just 4 degrees over Zero Celsius. I was sitting in my car with engine turned off, busy passively smoking while thinking furiously about something. My habit is rare these days. Still have to get used to aerodynamic cars. Also these long hot summers.

Scene is an alley behind a common supermarket in a very suburban area. To some casual observer, one side was a solid wall of concrete blocks painted some sick *beige* or salmon hue in glossy latex. No graffiti. Some pipes or ducts with steel doors at regular intervals, all locked. Some dumpsters in a front of a surface that was twenty feet tall and ran for at least 1,800 feet. There were no other cars visible on this bare, clean expanse of wet asphalt. No litter here. On the other side of this alley was a deep ditch full of foliage, some of it still green. Then across from that meadows and woods. Beyond that I could see a few *hirise* blocks. There were no other people present, which I was looking for. I had been sitting there in some trance for hours, which was foreign to my nature. You see, for days now my *celfone* has not been working but my regular one is still okay. I tried to get my relatives and/or friends to comment on this, but to no avail. Also I did not have the guts to ask any cops about this, nor strangers out on the street. Now it was really beginning to bug me. Not only the usual, but also regular TV on the airwaves. Cable was still okay. Why?

It had not rained all summer. Well almost. Now in the Fall of 2012 we have to have hurricanes in the South with chilly rain in the North. Well it has been raining for a solid week now. Why? Actually the climate was not of "issue" to me but this annoying problem with my expensive but useless *celfone* was getting on my nerves. What really got to me was that silence from other human beings. Across the stream, some movement in foliage. In that cold drizzle, fronds of weeping willow parted to reveal a short figure. This being was appeared to be an adult but petite, as they say. A perfect female figure. Totally nude and - this shocked me - of a milky blue skin color. Her face was childlike with a purple Sumerian haircut. Eyes orange changing to green while always flourescing from dull to shiny. Eyes were the giveaway. Just not human. Was her blood then made of freon which possibly consists of carbon, chlorine and flourine? Did this person get her energy from a diet of carbon and sulfur? The figure stepped out to reveal herself without shame. By instinct I could tell it would not be able to embrace me even though it was a very attractive female. She, or it, slowly came down the slope, crossed the shallow creek and then climbed up the grassy rise. Then it calmly walked up to my car. Showers continued as she stared at me in silence. As she stared without any expression, I heard a tiny voice far away say something like "Can I help you?"

Her lips parted slightly and she appeared to try to speak. Then a blur of some kind flashed over her face. She was masking her emotions. I told her that my Redberry was a piece of $500 junk. "Thanks for nothing." I said in a casual way. I had no intention of becoming angry, because this of all my possible causes for my problem was the last thing on my mind. For a random joke, consider what some dork in the 1950s would have suspected. In those simple days, our enemy was known.

The next "unit" of something from her mind was some abstruse joke about me being lucky that nobody (such as she herself) having accused me of somehow emitting "junk" on a verbal level. As the Occupy Rads say, "That's junk!" As you can tell, I am not one of them. So at that point I did not honestly know how to take their intervention. Did they do us a favor by stopping our Celfone Wars or not? It was so wishy washy, like most Post 90s polits.

My reverie was interrupted, but silently, by suddenly seeing four new figures appearing in the distance. They loomed over the puddles, slowly advancing down this alley in a vee formation. As they got closer I could see that all were dressed, but of the same Race. One wore a silver bikini. Since she had a full figure, the bottom covered her entire abdomen while her top covered her bags. It was a proper kind of military uniform. As we can assume. As you can readily imagine, I was a sucker for 1920s Hollywood. Or perhaps cheap novels from 1938. I mean, why say "Race" instead of "species". How gauche. Well, I am sorry but this is not a joke. It has been sad reality for many moons for me.

The other three (behind her as I digress) were a bit taller but of the same biome or sisterhood. Well. Okay. They had short hair and wore black teflon baggy pilot suits. Black combat boots. Carried assault rifles. Their, as I assume, Commander in her titanium fibre bikini stopped, spread her muscular legs in a wide stance and raised her left hand. I have no idea why but then she opened her mouth to speak gibberish. Her "soldateska" of three halted, spread out and then froze in formation. They pointed guns at all of three possible directions of threat as per some alien book of rules. Why the Hell not? I can imagine anything from these sweet weirdies. They did in no way amuse me. Up the alley, down the alley plus into the bushes. We were safe. They could empty magazines into all directions.

"I have a joke for you" I said. Leader asked what. Lips moved but no sounds came from them. It came directly from her brain. A weak, distant voice intoned: I cannot imagine. So I said, "Well if you were by some goofy chance to ask to be taken to my leader, then what would be the best answer?" These people cannot possibly be that smart. So I had to feed her punchline. It was very basically that I had to "take" her as my "leader" which is bad news for the United Nations, I guess. I already had no respect for my Prime Minister but that was so old it was a joke by the 1970s. Ha. Ha. So. Very.

She then beamed at me her own ancient joke of heritage, no doubt going back giga years. She once upon a time teleported herself into the vast Void again, as her hobby. She had a few minutes of reactive gas mixture in her lungs, and was beginning to feel the vacuum plus immense and orgasmic vertigo. It was fun. Then the Void boomed out "The universe does not exist!" Then she despaired in fear. But suddenly that same voice started to laugh as galaxies exploded out of nowhere to surround her. Then the same voice said "Ha, ha! I kid! Actually I do exist." Then she stared at me with a really dumb expression. Was that their idea of humor? Then the naked one faded away. After that the others also became mere phantoms, to finally all vanish totally. I ended up alone again. The rains continued.

END OF STORY

Afterword:

Upon some days after, I surmised that microwaves all over our globe have been affected. What that will that do to our society? We Men are clueless about this. Our hitech industry in the toilet. All things are possible in these times. Also consider that in 1978 certain other short beings failed to show up after a certain **flap** as they call it in the USA. Also recall that their ships **do** seem to work on heat created by intense microwaves. But that was a younger, weaker Race. I am talking of course about our famous **Grey Guys** who came in ships between 1947 and 1977, then vanished. Some of us suspect that **Archaea,** who can teleport and must be far more advanced; an older Race, did our small ones in. Both came from Afar like in some corny tome. That is possible, I figure to myself as I smoke in my car and think of those funny girls with their funny behavior. Who won? Who are they fighting and how? What does any massive organized conflict between stellar factions mean?

None of **them** have ever used words we know. I once heard some comments about how our nation was supposed to be on what was once called Atlantis, but is now North America. As for aliens, are they good or evil? Come to explore in peace or conquer? What does Peace and War mean to Them? Even so, how can We tell these days? Amen.

Story "Rain" written in 2012.

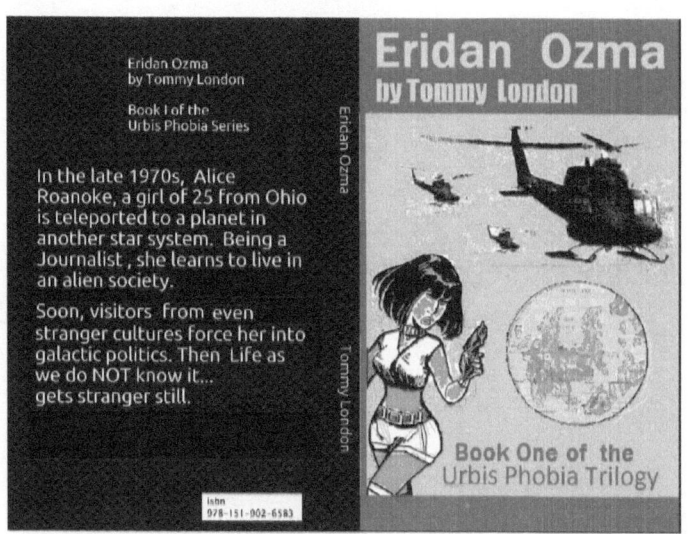

Below: Short story "Fear City" which started this entire trilogy. Created in 1996. Like a pilot for a TV series. Turned into my roman, Eridan Ozma.

Story: Fear City

This appeared in 1996 as pilot episode for Book One.

We spend much time these days wondering how and why Mankind has survived so far, an for good reason: The fact that I may have contributed to that process without knowing about it. It started with my car running out of gas at 04:00am on some deserted hiway in the suburbs of Cleveland, Ohio. Having no cash nor credit card, I had to phone my friends to come over and rescue me. I stood in a hot, brightly lit phone booth on July 20th, 1978. A garishly lit modern gas station stood a few dozens of feet away. Behind huge windows I could see desks, shelves, cash registers, posters, stacks of batteries, ziggurats of oil cans and one attendant. Also a small snack bar with pop machines, food bins and so forth. It was all lit up by glaring white neon, with a vast expanse of black woods beyond. Overhead hung the clear, bright Milky Way as if it were important to know which galaxy we live in.

[**Editor:** Sorry for my overbearing Gernsback & Campbell basic offering of comfort, but it really does matter which galaxy we do live in, Like, when the first saucers contacted the First Contactee, as in say, 1947, they invariably claimed to be "From some other galaxy." It was in any case, some other Star System. As if some naive Earthian did know astronomy any better. And such things did matter. Said aliens originated from Epsilon Eridani.]

Sweat moistened my Teeshirt and bra. The salty liquid ran down my legs. My jeans got soaked. They were tight, unflared but still faded properly. In the distance my 1958 Dodge sat there as a dark, hulking beast of steel. My random call was making me feel guilty. I even thought of offering something "good" to the station attendant for some gas... like my body. The phone rang forever at the other end. A certain piece hidden in my Dodge came to mind but I did not intend to cowboy this poor guy for a few lousy gallons of gasoline. Then my line was suddenly cut off. I became dizzy and felt my body sink down wards. My skin grew cold and tingled in perverse manner. What a substantial being I am, I thought. As my eyes closed a star field rushed upward to expand into my face. The last thing I recall was a buzzing noise as I became a dead weight and felt as if I were sinking into the ground.

[pause]

I recall waking up in a strange room. Well, there are in fact, many strange rooms in Cleveland that I have awoken in. But not this one! I could tell at once this was a totally alien place. The room itself was nothing but a large expanse paved with lime green shag carpet, white walls and a low ceiling. One half of my suite was a platform under the carpet, with a curve for a step. The walls also had curved corners, which disoriented the human eye. All of this made for odd effects.

The main room was forty by thirty feet, and the whole wide side was one long window of thick glass. The glass was tinted orange and revealed a strange skyline half hidden by smog. The lower half of the room had two side doors of aluminum, which slid in and out of the wall automatically by a sensor device. Close to the window was a modern kitchen made of aluminum, brass and stainless steel. The one in the back was a bathroom made mostly of lime green marble, featuring a sunken tub. Both rooms were a foot lower than the main one to prevent flooding.

[pause]

Forgive my archeology tour. When I had gotten off the floor, I noticed that I was soaked with perspiration, leading me to assume that I had been teleported here. I drank water from the kitchen tap gladly because the room was hot. The air was stale, but a push on a button gave me cool fresh air. The vast skyline outside was really impressive under its grey overcast.

After a long time spent exploring the hallways I concluded that I was in a totally vacant but functional apartment complex about 60 stories tall. All doors were locked. I took an elevator down to the lobby and found that it looked like the same we have on

Terra. However, this front door was broken glass. I stepped out to emerge into a grey haze. It was even hotter here with tropical foliage in their garden. Even so, I could tell this area was one gigantic cluster of similar hirises, which went on forever into the distance.

Suddenly, right in front of me palmetto fronds parted to let a small female approach me. Her skin was light blue and her hair a darker shade of the same hue. She wore a skimpy, fragile outfit of many colors and her smooth features made her look even more fragile. She was only four feet tall and wore sandals. As she stared back at me, she sneered and pointed to my clunky cowboy boots. My clothing was also too sturdy for her. I waved her over. My Dodge and Colt 38 had been left behind by whatever cosmic forces had stranded me here. Combined with my size and predatory instincts, both objects would be lethal weapons here.

After a while my forehead grew chilly. A small blue dot came into my field of vision. Then the word "Telum" was audible. And then, knowing I was on the right track, I pointed at the big intense sun up above and got the words "Epsilon Eridani" as a reply. This was not vocalized but done in telepathy. She then indicated the place I had just exited and invited me in with a wave. She resided and worked here as a concierge and appeared to be the only person left. Being the de facto owner, she allowed me to play with the PC in the rental office. I've read cheap novels about the idea of a future dystopia. They assume that the so called Computer Revolution of the next decade will destroy our moral values. Well, I do not buy that theory at all. Yet, I am very much against Social Darwinism. My long hair has nothing to do with it. Yet I was to find out later that the Eridan were more advanced technologically than we were in 1978. Despite huge areas of vacant housing, their society still had activity of some kind. There were people and cars moving about down there. Even here in the suburbs.

My respect for my new friend's candle power came soon. We were working on her PC when I smiled and then patted her shoulders. Then showed her a US copper penny from my pockets, to which she responded with a smile of her own. "Copper," her mind said "therefore green or blue blood." Then she pulled out filing drawers and showed me moist, rotting paper records naming her former tenants. There was alien script all over the place but I did not bother trying to learn it. For Etiana, which was her name; had been teaching me telepathy.

Note:
[So ends this sample of "Fear City" the story of June, 1996. This one inspired our Trilogy of 1997 now called Urbis Phobia. In the later, expanded version, which runs 170 pages, some vital issues have been changed. Above, Alice gets to wander around her Superbloc. Her own door was unlocked. In the book, she is trapped and has to explore an outside world by astral travel. A talent she has only read about. Just another modern American Fantasy Fan. How Seventies! This whole setup was by Zeno, who has agents watching both Alice and Etiana. Well, Etiana is only eleven *Xiotan* years old, hence 9.04 Earth years. Yet this minor works as the tower Super. (Zeno owns

this block along with many other real estate in Gulf City.) She was hired by Zeno to spy on her human prisoner. Unemployment being extreme here, the law says one has to be 18 to work. More to the point, other than secular concerns, Alice has to learn to send out an astral body. or soul. Then, she also has to levitate and even teleport her organic body at will. Those talents are known to Zeno and a small elite of her Race. This Cosmic Elite are masters of the sidhii.]

[One detail: Alice is not smart. Had she been, she would have deduced that her car and gun were parked safe and sound in the garage of her own block. Superbloc 334.]

THE END

Gulf City as alien paradise. A day on the beach for Alice.

Publisher notes:
Paperback Edition of "FTL Paper" which was adapted from our short essay written in November of 1983. Art work & photos were added much later. This Amazon Inc. edition is mix of both scientific data as theory, and fiction. Some of this may be for mature readers only.

Published in 2015.
Isbn: 978-197-355-8224

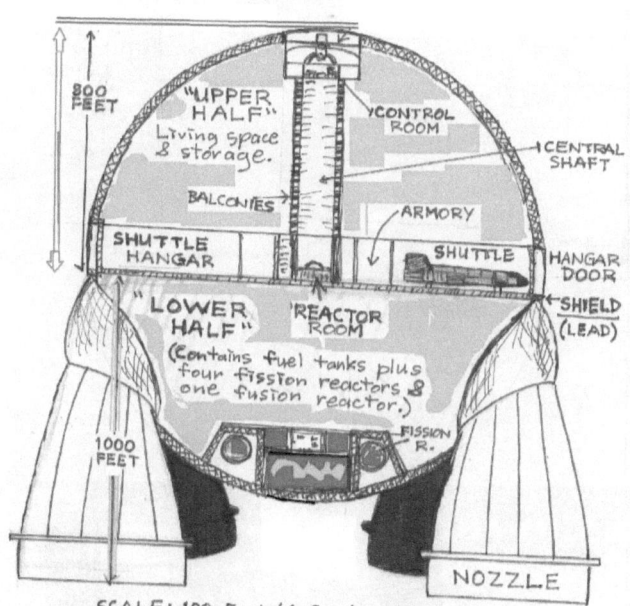

SCALE: 100 Feet / 1 Centimeter

Total length: 1800 feet. Diameter of globe: 1400 feet.
Central shaft: 100 feet wide by 650 feet deep.
Three shuttles in three hangars (exit by side doors).
Engines: Four fusion reactors (plutonium) powering one
fusion reactor to produce plasma; also capable of using
"antimatter" fuel. Electronics by ABM Inc.

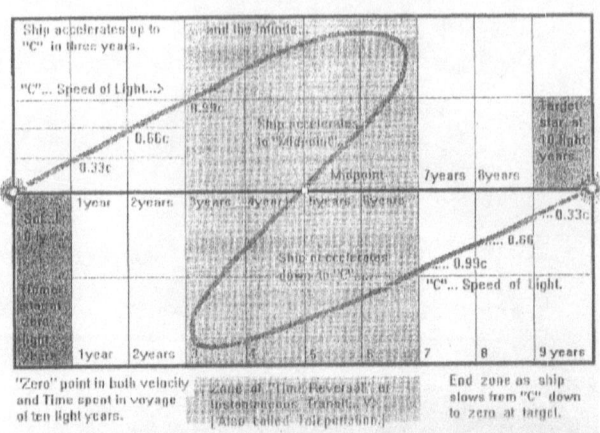

"Zero" point in both velocity and Time spent in voyage of ten light years.   Zone of "Time Reversal" or Instantaneous Transits. (Also called Teleportation.)   End zone as ship slows from "C" down to zero at target.

Donovan Mu

Essay by Tomas London

Essays about ancient and alien methods for creating cartography for us

ASTRONAUT GODS OF THE MAYA

Extraterrestrial Technologies in the Temples and Sculptures

ERICH von DÄNIKEN

Ree City
the Planetary
Capital of Cetiwan
in Tau Ceti System

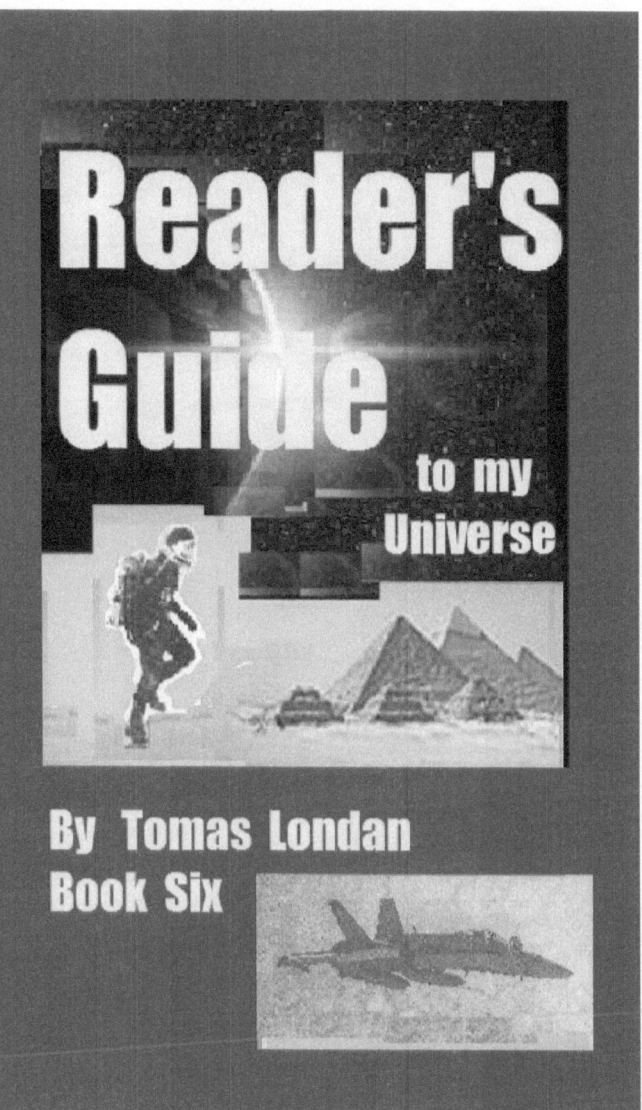

NOTE:
The above art was some of my own books, both fact and fiction plus three books from other authors we recommend. All deal more or less with the issue of Exobiology, which has been covered in detail by the MUFON group, which we had once been a member of. Also do see the Ancient Aliens TV show, Season Seven. For their recent meeting in Congress, Washington, D.C.

www.ingramcontent.com/pod-product-compliance
Lightning Source LLC
Chambersburg PA
CBHW031418210526
45464CB00005B/1944